MY FARM.

MY FARM

OF EDGEWOOD:

A Country Book.

BY THE AUTHOR OF

"REVERIES OF A BACHELOR."

——*"it was all grown over with thorns, and nettles covered the face thereof, and the stone-wall thereof was broken down. Then I saw and considered it well: I looked upon it, and received instruction."* Proverbs xxiv, 30.

NEW-YORK:
CHARLES SCRIBNER.
1863.

JOHN F. TROW,
PRINTER, STEREOTYPER, AND ELECTROTYPER,
50 Greene Street, New York.

TO

MY OLD FRIEND,

GEN. WILLIAM WILLIAMS,

OF NORWICH,

IN GRATEFUL RECOGNITION OF HIS MANY KINDNESSES,

DATING FROM THE TIME

WHEN HE AIDED ME IN MY FIRST CARE OF A NEW ENGLAND FARM;

AND IN TOKEN OF MY RESPECT FOR HIS WORTH,

I DEDICATE

THIS TRANSCRIPT OF ANOTHER AND LATER

FARM EXPERIENCE.

PREFACE.

A FRIEND asks,—"Are you not tired, then, of that fancy of Farming? Is it not an expensive amusement; is it not a stupefying business?

"Do you find your brain taking breadth or color out of Carrot-raising, or Pumpkins? Poultry is a pretty thing, between Tumblers, and Muscovy ducks; but can you not buy cheaper than you raise,—without the fret of foxes and vermin,—in any city market?

"Shall I sell out and join you? Shall I teach this boy of mine (you know his physique and that gray eye of nis, looking after some eidolon) to love the country—so far as to plant himself there, and grow into the trade of farming? A victory over the forces of nature, and of the seasons,—compelling them to abundance,—is no doubt large; but is not a victory over the forces of mind, which can only come out of sharp contact with the world, immenſely larger?"

In my reply,—loving the country as I do, and wiſhing to set forth its praiſes; and believing as I do, in the God-appointed duty of working land to its top limit of producing power,—I said a great deal that looked like a mild Georgic.

And yet, with a feeling for his poor boy, and a remembrance of what crisp salads I had found in the city markets, after mine were all mined and devoured by the field-mice,—I wrote a great deal that had the twang of Melibœus in the eclogue,

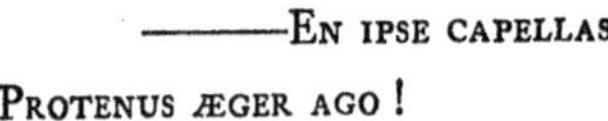

———En ipse capellas
Protenus æger ago!

In short, in my reply, I beat about the buſh:—so much about the buſh in fact, that this book came of it.

Edgewood, 1863.

CONTENTS.

IV.—*HINDRANCES AND HELPS.*

THE SEARCH AND FINDING.

I.

THE SEARCH AND FINDING.

IT was in June, 18—, that, weary of a somewhat long and vagabond homelessness, during which I had tossed some half a dozen times across the Atlantic,—partly from health-seeking, in part out of pure vagrancy, and partly (*me tædet meminisse*) upon official errand—I determined to seek the quiet of a homestead.

There were tender memories of old farm days in my mind; and these were kindled to a fresh exuberance and lustiness by the recent hospitalities of a green English home, with its banks of laurestena, its broad-leaved rhododendrons, and its careless wealth of primroses. Of course the decision was for the country; and I had no sooner scented the land, after the always dismal sail across the fog banks of Georges' shoal, than I drew up an advertisement for the morning papers, running, so nearly as I can recall it, thus:—

"Wanted—A Farm, of not less than one hundred acres, and within three hours of the city. It must have a running stream, a southern or eastern slope, not less than twenty acres in wood, and a water view."

To this skeleton shape, it was easy, with only a moderately active fancy, to supply the details of a charming country home. Indeed, no kind of farm work is more engaging, as I am led to believe, than the imaginative labor of filling out a pleasant rural picture, where the meadows are all lush with verdure, the brooks murmuring with a contented babble, cattle lazily grouped, that need no care, and flowers opening, that know no culture. This kind of farm work is not, to be sure, very profitable; and yet, as compared with a great deal of the gentleman-farming which I have had occasion to observe, I should not regard it as extravagant. Perhaps it would not be rash to put down here some of the pictures which I conjured out of the advertisement.

At times, it seemed to me that an answer might come from some Arcadia lying upon the cove banks of an inland river: the cove so land-bound as to seem like a bit of Loch Lomond, into which the north shores sunk with an easy slope, whose green turf reached to the margin, and was spotted here and there with old and mossy orcharding; the west shore rose in a stiff bluff that was packed close with hemlocks and maples; while beyond the bluff a rattling stream came down over mill dykes and through swift sluices, and sent its whirling bubbles far out into the bosom of the little bay. West of the bluff lay the level farm lands; and northward of the green slope which formed the northern shore, it seemed to me that wooded hills would rise steep and ragged, with such wildness in them as would make admirable setting for the sloping grass land below, and the Sunday quiet of the cove. It seemed to me that possibly there might be an oyster bed planted along the shore, which would help out the salads that should be planted above; and, possibly, a miniature dock might be thrust out into the water, at which some little pinnace might float, with a gay pennant at her truck.

Possibly it does; possibly there is such a place; but for me it was only a picture.

Again, it seemed to me that the farm house would nestle in some little glen upon the banks of a river, where every day crowded boats passed, surging up against the current; or gliding down with a meteor-like swiftness.

In this case, the slopes were many: a slope eastward from the house door to the banks of a little brook that came sauntering leisurely out from the wood, at the bottom of the glen; a slope from the house up to the hills piling westward; slopes on either margin of the glen; and above them, upon higher ground, pasture lands dotted with stately trees; while a fat meadow seemed to lie by the river bank, where the little brook came sauntering in. There, and thereabout, whisking their sides, stood the cattle, as in a Flemish picture—as true, as still, and just as real. There may be such cattle whisking their tails, but they are none of mine.

Then, it seemed the home should be upon an island, looking down and off to the sea, where ships shortened sail, and bore up for the channel buoys, which lay bobbing on the water. There, the farm land ended in a pebbly beach, on which should lie a great drift of sea weed after every southeaster. The wood was a stately grove of oaks, taking the brunt of the northwesters that roared around the house in autumn, and making grateful lee for the pigeons that dashed in and around the gables of the barn. The brook seemed here a mere creek, which at high water should be flooded, even with the banks of sedge; and when the tide was out, showed half a dozen gushing springs which plied their work jantily till the ebb came, and then, after coquetting and toying with their lover, the sea—were lost in his embrace.

——Only a fancy! If there be such a lookout from farm windows, the ships come and go without my knowledge; and the springs gush, and die in the flow of the tide, unknown to me.

Again, it seemed that answer would come from some remote valley side, away from the great highways of travel, where neither sail nor steamer obtruded on the eye;—where indeed a sight of the sea only came to one who climbed the tallest of the hills which sheltered the valley. Half down the hills an old farm house, with mossy porch, seemed to rest upon a shelf of the land. A cackling, self-satisfied, eager brood of fowls were in a party-colored cloud about the big barn doors; a burly mastiff loitered in the sun by the house steps, mild-eyed cows were feeding beyond the pasture gate; a brook that was half a river, came sweeping down the meadows in full sight—curving and turning upon itself, and fretting over bits of stony bottom, and loitering in deep places under alluvial banks, where I knew trout must lie—then losing itself, upon the rim of the farm, in tangled swamp lands; where, in autumn, I knew, if the farm should be mine, I could see the maples all turned into feathery plumes of crimson. But I did not; plumes of crimson I see indeed each autumn, but they are at my door; and a great reach of water comes streaming to my eye without lifting from my chair.

It was not from mere caprice that my advertisement had been worded as it was. For the mere establishment of a country home, one hundred acres might seem an unnecessarily large number, as indeed it is. But I must confess to having felt an anxiety to test the question, as to whether a country liver was really made the poorer by all the acres he possessed beyond the one or two immediately about his homestead. Indeed I may say that I felt a somewhat enthusiastic curiosity to know, and to determine by actual experiment, if farm lands were simply a cost and an annoyance to any one who would not wholly forswear books, enter the mud trenches valorously, and take the pig by the ears, with his own hands.

A half dozen acres, which a man looks after in the intervals of other business, and sets thick with his fancies, in the shape of orchard houses, or dwarf pear trees, or glazed graperies, offer no solution. All this is in most instances, only the expression of an individualism of taste, entered upon with no thought of those economies which Xenophon has illustrated in his treatise, and worse than useless as a guide to any one who would make a profession of agricultural pursuits.

With fifty or a hundred acres, however, steaming under the plough, and with crops opening successively into waving fields of green,—into feathery blossom,—into full maturity; too large for waste; too considerable for home consumption; enough, in short, to be brought to that last test of profit—a market, and a price; then the culture and its costs have a plain story to tell. The basis will not be wanting for an intelligent decision of the question—whether a man is richer in the cultivation of a hundred acres, or of ten; whether, in short, farming is a mere gross employment, remunerative, like other manual trades, to those immediately concerned; or whether it is a pursuit subject to the rules of an intelligent direction, and will pay the cost of such direction, without every-day occupancy of the field.

My advertisement named three hours' distance from the city, as one not to be exceeded. Three hours in our time means eighty miles; beyond that distance from a great city, one may be out of the eddies of its influence; within it, if upon the line of some connecting railway, he is fairly in a suburb. Three hours to come, and three to go, if the necessity arise, leave four hours of the pith of the day, and of its best sunshine, for the usurers of the town. Double four hours of distance, and you have a journey that is exhausting and fatiguing; double two hours, or less, and you have an ease of transit that leads into temptation. If a man then honestly determines to be a country liver, I hardly know a happier mean of distance than three hours from the city. If, indeed, he enters upon that ambiguous mode of life which is neither city nor country, which knows of gardens only in the night time, and takes all its sunshine from the pavements, which flits between the two without tasting the full zest of either—of course, for this mode of life, three hours is too great a distance. The man who is content to live in grooves on which he is shot back and forth year after year—the merest shuttle of a commuter,—will naturally be anxious to make the grooves short, and the commutation small.

I bespoke in my advertisement no less than twenty acres of woodland. The days of wood fires are not utterly gone ; as long as I live, they never will be gone. Coal indeed may have its uses in the furnace which takes off the sharp edge of winter from the whole interior of the house, and keeps up a night and day struggle with Boreas for the mastery. Coal may belong in the kitchens of winter ; I do not say nay to this : but I do say that a country home without some one open chimney, around which in time of winter twilight, when snows are beating against the panes, the family may gather, and watch the fire flashing and crackling and flaming and waving, until the girls clap their hands, and the boys shout, in a kind of exultant thankfulness, is not worthy the name.

And if such a fiery thanksgiving is to crackle out

its praises—why not from a man's own ground? There is no farmer but feels a commendable pride in feeding his own grain, in luxuriating upon his own poultry, in consuming his own hay—why not burn wood of his own growing? It is not an extravagant crop. Thirty years of rocky, wild land, else unserviceable, will mature a good fire-crop; and if there be chestnut growth, will ensure sufficient size for farm repairs and fencing material. A half acre of average growth will supply at least one roaring winter's fire, beside the chestnut for farm purposes. And thus with twenty acres of wood, cut over each year, half acre by half acre, I have forty years for harvesting my crop; and then, the point where I entered upon my wood field is more than ready for the axe again. Indeed, considering that thirty years are ample for the growth of good-sized fire wood, I have a margin of ten years' extra growth, which may go to pin money; or may be credited to some few favorite timber trees that stand upon the edge of the pasture, and pay rental in the picture they give of patriarchal grace,—to say nothing of an annual harvest of chestnuts.

Woodland, again, gives dignity to a country place; it shows a crop that wants a man's age to ripen it; a company of hoary elders—conservatives, if you will—to preside amid the lesser harvests, and to parry the rage of tempests. Mosses plant their white blight, as gray hairs come to a man; but the core is sound, and the life sap swift, and in it are the juices of a thousand leaves.

A wood, too, for a contemplative mind is always suggestive. Its aisles swarm with memories; the sighing of the boughs in the wind brings a tender murmur from the farthest days of childhood, when leaves rustled all the long summer at the nurse's window. Bird-nesting boyhood comes again to sit astride the limbs—to hunt for slippery elm, or the fragrant leaves of young wintergreen, or the aromatic roots of sassafras.

This scarred bole, so straight and true, reminds of still larger ones in the forest of Fontainebleau; the chestnuts recall the broad-leaved ones of the Apennines; the hemlocks bring to memory the kindred *sapin* of the Juras, under whose shade I sat upon an August day, years ago, panting with the heat, and looking off upon the yellow plains which stretch beyond the old French town of Poligny, and upon the shadows of clouds, that flitted over the far and "golden sided" Burgundy.

Next, the coveted place was to have its quota of running water. It would be a very absurd thing to go far to find reasons for the love of a brook. There are practical ones of which every farmer knows the force; and of which every farmer's boy, who has ever driven a cow to water, or wet a line in the eddies, could be exponent.

And in the romantic aspect of the matter, I believe there is nothing in nature which so enlaces one's love for the country, and binds it with willing fetters, as the silver meshes of a brook. Not for its beauty only, but for its changes; it is the warbler; it is the silent muser; it is the loiterer; it is the noisy brawler; and like all brawlers beats itself into angry foam; and turns in the eddies demurely penitent, and runs away to sulk under the bush. A brook, too, piques terribly a man's audacity, if he have any eye for landscape gardening. It seems so manageable, in all its wildness. Here in the glen a bit of dam will give a white gush of waterfall, and a pouring sluice to some overshot wheel; and the wheel shall have its connecting shaft and whirl of labors. Of course there shall be a little scape-way for the trout to pass up and down; a rustic bridge shall spring across somewhere below, and the stream shall be coaxed into loitering where you will—under the roots of a beech that leans over the water—into a broad pool of the pasture close, where the cattle may cool themselves in August. In short, it is easy to see how a brook may be held in leash, and made to play the wanton for you, summer after summer. I do not forget that poor Shenstone ruined himself by his coquetries with the trees and brooks at Leasowes. I commend the story of the bankrupt poet to those who are about laying out country places.

Meantime our eye shall run where the brooks are running—to the sea. It must be admitted that a sea view gives the final and the kingly grace to a country home. A lake view and a river view are well in their way, but the hills hem them; the great reach which is a type, and as it were, a vision of the future, does not belong to them. There is none of that joyous strain to the eye in looking on them which a sea view provokes. The ocean seems to absorb all narrowness, and tides it away, and dashes it into yeasty multiple of its own illimitable width. A man may be small by birth, but he cannot grow smaller with the sea always in his eye.

It is a bond with other worlds and people: the sail you watch has come from Biscay; yesterday it was white for the eye of a Biscayan; your sympathies touch by the glitter of a sail.

The raft of smoke drifting from some steamer in the offing is as humanizing, though it be ten miles away, as the rattle of your neigbor's wagon by the door.

You live near a highroad to take off the edge from loneliness and isolation; but a travelled sea, where all day long white specks come and go, is the highway of the world; and though you do not see these neighbors' faces, or catch their words, the drifted vapor, and the sheen of the sails, and the streaming pennants yield a sense of nearness and companionship that gives rein and verge to a man's humanity.

Then, physically,—what reach! Heaven and earth touch their great circles in your eye; the touch that bounds human vision wherever you may go. No height can lift you to a grander touch, or alter one iota its magnificent proportions. With a land horizon, it may be an occasional hill that conceals the outmost bound,—a temple or a tree; it is various and uncertain; even upon the prairie a harvest of flowers may fringe it with an edge that the autumn fires consume, or which a trampling herd may beat down; but where sea touches sky, there, forever, is the line immutable, which runs between our home and the spacious heaven, that buoys, and bears us. And thence, with every noontide, the sun pours a fiery profusion of gold up to your feet; and there, every full moon paves a broad path with silver.

So, with each of the features I have claimed, come kingly pictures;—not least of all to the gentle slope south or eastward, which should catch the first beams of the morning, and the first warmth of every recurring spring.

In a mere economic point of view, such slope is commended in every northern latitude by the best of agricultural reasons. In all temperate zones two hours of morning are worth three of the afternoon. I do not know an old author upon husbandry who does not affirm my choice, with respect to all temperate regions. If this be true of European countries, it must be doubly true of America, where the most trying winds for fruits, or for frail tempers, drive from the northwest.

And with the slope, as with the wood and with the sea, come visions;—visions of sloping shores of bays, into whose waters the land dips with every recurring tide; and where, as the gentlest of tides fall (so upon the Adriatic coast), an empurpled line of fine sea mosses lies crimped upon white sand, and pearly shells glitter in the sun. Or,—of lake shores, gentle as Idyls (so of Windermere), with grassy slopes so near and neighborly to the water, that the mower, as he clips the last sentinels in green, sweeps his blade with a bubbling swirl of sound, quite into the margin of the lake.

Southern slopes, again, suggest luscious ripeness. The first figs I ever gathered, were gathered on such a slope in a dreamy atmosphere of Southern France, with the blue of the Mediterranean in reach of the eye, and the sweetest roses of Provence lending a balmy fragrance to the air.

Sheltered slopes recall too, always, what is most captivating in rural life. You never see them or look for them even, in Dutch-land—in Poland, never; in Prussia, or on the highways of travel in France, never. And no rural poems, or pictures that haunt the memory, were ever rhymed or sketched in those regions. Theocritus lived where lie the sweetest of valleys; Tibullus and Horace both knew the purple shadows that lay in the clefts of the Latian hills. Delille chased his rural phantoms beyond the Burgundian mountains, before they had taken their best form.

But in the English Isle—by Abergavenny, by Merthyr, under the Tors of Derbyshire, in the lea of the Dartmoor hills,—abreast of Snowdon—what sheltered greenness and bloom! What nestling homesteads!

I must not forget to give a sequence to my story. I had entered my advertisement. Was it possible that any one in the possession of such a place as I had roughly indicated, would be willing to sell?

For twenty-four hours I was in a state of doubt; after that time, I may say the doubt was removed. I must frankly confess that I was astounded to find what a number of persons, counting not by tens, but by fifties, and even hundreds, were anxious to dispose of a "situation in the country" which fully corresponded to my wishes (as advertised.)

Were the people mad, that they showed such eagerness to divest themselves of charming places? Or were my fine pictures possibly overdrawn? And yet, who could gainsay them; are not trees, trees—and brooks, brooks—and the sea, always itself?

I think my New York friend, to whom I had ordered all replies to be addressed, may have handed me a peck of letters;—blue letters, square letters, triangular letters, pink letters (in female hand), and soberly brown letters.

It was a mortification to me to reflect that so many fine places should be thrown upon the market at the first hint of a purchaser; "places most convenient;" places on a "lovely shore;" places by rivers; places with commanding views; places on prospective railways; places innumerable.

Not a few of the propositions contained in these letters were, at first sight, plainly inadmissible; as where a sanguine gentleman suggested that I should make a slight change of programme, so far as to plant myself on the shores of the Great Lakes, or in a pretty retiracy, among the fine forests along the Erie railroad.

Another, "in case I found nothing to suit elsewhere," could recommend "a small place of ten acres, in a thriving country town, two minutes' walk from the post office, house forty by thirty-five, and ten feet between joints, stages passing the door three times a day, large apple trees in the yard newly grafted, and the good will of a small grocery, upon the corner, to be sold, if desired, *with the goods*, and healthy."

Inadmissible, of course; and the letter passed over into the hat of my friend. Another letter, from a widow lady, invited attention to the admired place of her late husband: he had "an unusual taste for country life, and had expended large sums in beautifying the farm; marble mantels throughout the house, Gothic porticos, and some statuary about the grounds. There was a gardener's cottage, and a farmer's house, as well as another small tenement for an under-gardener, and twenty acres of land, of which six in shrubbery and lawns." The architecture seemed to me rather disproportionate to the land; inadmissible upon the whole, as a desirable place on which to test the economies of a quiet farm-life.

I can conceive of nothing so shocking to a hearty lover of the country, as to live in the glare of another man's architectural taste. In the city or the town there are conventional laws of building, established by custom, and by limitations of space, to which all must in a large measure conform; but with the width of broad acres around one, I should chafe as much at living in the pretentious house of another man's ordering and building, as I should chafe at living in another man's coat. Country architecture, whose simplicity or rudeness is so far subordinated to the main features of the landscape as not to provoke special mention, may be of any man's building; but wherever the house becomes the salient feature of the place, and challenges criticism by an engrossing importance as compared with its rural surroundings, then it must be in agreement with the tastes and character of the occupant, or it is a pretentious falsehood.

Perhaps I ought to beg the reader's pardon for this interpolation here, of a law of adjustment in respect to the country and country houses, which would have more perfect place in what I may have to say upon the general subject of rural architecture.

At present I return to my stock of pleasant advisory letters:

A tasteful gentleman, of active habits, calls my attention to a park of which he is the projector, and within which several desirable places, with admirable views, remain unsold; while land in the neighborhood might be secured at a reasonable valuation, for such farm experiments as I might be tempted to enter upon. Attention is particularly called to the social advantages of such a neighborhood, where none but gentlemen of character would be permitted to purchase, and where the refinements of city intercourse would be, &c., &c.

Now it so happens that I never heard of a park upon this mutual method, where there did not arise within a few years a smart quarrel between two or more of the refined occupants. The cows, or the goats, or the adjustment of water privileges, are sure to form the bases of noisy differences, in the management of which, I am sorry to say, the amenities of the town are not greatly superior to the amenities of the country. Aside from this danger, I have not much faith in the marketable coherence of those rural tastes which would belong to a promiscuous circle of buyers. A community of cooks, or of coal-heavers, I can conceive of, but a community of ruralists, or of amateur farmers, quite passes my comprehension. I say amateur farming, for I know of no farming which is so amatory in the beginning, and so damnatory in the end, as that which delights in a suburban house, and in a sufficient quantity of ground a few miles away, where, under the wary eye of some sagacious Dutchman or Irishman, the cows are to be fed, the weeds pulled, the chickens plucked, and the new industry and profit developed generally. It is very much as if a man were to enter upon the business of whaling by taking rooms at the Pequod House, and negotiating with some enterprising skipper to tow a few tame whales into harbor, to be slashed up, and tried, and put into clean casks, on some mild afternoon of June.

In the latter case, we should probably have the oil and the bone; and in the other, we should perhaps have the butter and the eggs; in both, we certainly should have the bills to pay.

If a man would enter upon country life in earnest, and test thoroughly its aptitudes and royalties, he must not toy with it at a town distance; he must brush the dews away with his own feet. He must bring the front of his head to the business, and not the back side of it; or, as Cato put the same matter to the Romans, near two thousand years ago, *Frons occipitio prior est.*

But while I was thus compelled to discard certain propositions at their first suggestion, there were others which wore such a roseate hue as challenged scrutiny and compelled a visit. Thus, a very straightforward and business-like letter from a Wall-street agent informed me that his esteemed client, Mr. Van Heine, "was willing to dispose of a considerable country property thirty miles from the city, in a favorable location. The house was not large or expensive, possibly not extensive enough; there was old wood upon the place, the surface charmingly diversified, and in addition to other requisites, it possessed a mill site, mill, and small body of water, which, in the hands of taste, he had no doubt," &c., &c.

The agent regretted that he could give me no definite information in regard to the exact size of the property, or terms of sale, but begged me to pay a visit to the place before deciding.

The description, though not particularly definite, was yet sufficiently piquant and suggestive to induce me to comply with the hint of the agent. I liked the man's nomenclature—"a considerable country property;" it conveyed an impression of dignified quiet and retirement. The dwelling was probably a modest farmhouse, grown mossy under the shade of the old wood; possibly some Dutch affair of stone, with Van Heine gables, which it would be hardly decorous to pull down. I might add a little to its size, and so make it habitable; or, if well placed, it might—who knew—be turned into a cottage for the miller. There remained, after all this agreeable coloring, the small body of water and the diversified surface, which were enough in themselves to form the outlines of a very captivating picture.

I determined to pay Mr. Van Heine a visit. Obtaining all needed information from his agent, in regard to the locality and its approaches from the city, I set off upon a charming morning of June by one of the northern railways, and after an hour's ride, was put down at a station some five miles distant from the property. I drove across the country at a leisurely pace, stopping here and there upon a hilltop to admire the far-off views, and speculating upon possible improvements that might be made in the badly conditioned road. The neighborhood was not populous: indeed, it was only after having measured, as I fancied, the fifth mile, that I for the first time saw a party from whom I might ask special directions. I may describe this party as a tall man in red beard and red fur cap, with a black-stemmed porcelain pipe in his mouth, and pantaloons thrust into stout cowhide boots. He was striding forward in the same direction with me, and at nearly an equal pace.

"Did he possibly know of a Mr. Van Heine in this region?"

"Yah—yah," and the man, who may have been an emigrant of only four or five years of American nationality, pointed toward himself with a pleased and grim complacency.

"This was Mr. Van Heine, then, who has a country property to sell?"

"Yah—yah," and his smile has now grown eager and familiar.

His place is a little farther; and I ask him to a seat beside me.

"It is a farm he has to sell?"

"Yah—yah, farm."

I ask if the view is good.

"Yah—view—yah."

I venture a question in regard to the mill.

"Yah—mill—yah."

"Grist mill?" I ask.

"Yah—mill."

"For sawing?" I add, thinking possibly he might misunderstand me.

"Yah—sawing."

I venture to ask after his crops.

"Crops—yah."

The conversation was not satisfactory: we were driving along a dusty highway, and had entered upon a sombre valley, where there was no sign of cultivation, and where the only dwelling to be seen, was one of those excessively new houses of matched boards, perched immediately upon the side of the high-road, and with its pert and rectangular "joinery" offending every rural sentiment that might have grown out of the blithe atmosphere and the morning drive.

"Dish is de place," said my friend of the red beard and porcelain pipe; and I could not doubt it; there was a poetic agreement between man and house; but the mill remained—where was the mill?

Van Heine was only too happy: across the way—only at a distance of a few rods, not removed from the dust of the high-road, was the mill, and the "body of water." The new scars in the hillsides, from which the earth had been taken to dam the brook, were odiously apparent: but the investment had clearly not proven a profitable one: the capacity of the brook had been measured at its winter stage; even now, the millpond at its upper end showed a broad, slimy flat, which was alive with frogs and mudpouts. A few scattered clumps of dead and seared alders broke the level, and a dozen or more of tall and limbless trees that had been drowned by the new lake, rose stragglingly from the water—making, with the dead bushes, and the loneliness of the place, a skeleton and ghostly assemblage.

Mr. Van Heine had newly filled his pipe, and was puffing amiably, as I stood looking at the property, and at the sandy hills which rolled up from the further side of the pond, tufted with here and there a spreading juniper. The whole aspect of the property was so curiously and amazingly repugnant to all the rural fancies I had ever entertained, whether æsthetic, or purely agricultural, that I was excessively interested. My red-bearded entertainer clearly saw as much, and with violent and persuasive puffs at his porcelain pipe, and occasional iterative "dams" in his talk, (which had very likely sprung of unpleasant familiarity with the dam actual) he became explosively demonstrative and earnest.

I hinted at the shortness of the water; there was no denial on his part; on the contrary, frank avowal.

"Yah—dam—short," said he; "dat ish—enough for der farm—yah; but for der mill—dam—nichts" (puff).

I spoke in an apologetic way of the advertisement, and of certain requisites insisted upon; he had perhaps seen it?

"Advertisement—yah (puff)—yah."

I hinted at the slope.

"Yah—der slope."

"The slope to the south?"

"Oh yah—south (puff)—yah."

I explained by a little interpolation of his own tongue.

"Dam—yah—dis ish it; der is de pond; dish is south; dat ish der slope—to der pond—dam—yah."

"And the lands opposite?"

"Oh, dat ish not mine; der mill, der house, der pond, der land, vat you call der slope—dis ish mine."

I suggested the mention of a water view in the advertisement.

"View," said my red-bearded friend; "vat you call view?"

I explained as I could, teutonically.

"Dam! der vater view! (with emphasis); dis ish it; der pond, ish it no vater?—hein!—dam (puff)."

Even now I look back with a good deal of self-applause upon my success in extricating myself from the merciless and magnetic earnestness of the red-bearded Mr. Van Heine; I think of my escape from the dusty high-road, the angular joinery of the house, the bloated hills, blotched with junipers, the straggling trunks of the drowned trees, and the imperturbable insistance of the German, with his expletive dam and his black-stemmed porcelain pipe, as I think of escapes from some threatening pestilence.

Another country place was brought to my attention, under circumstances that forbade any doubt of its positive attractions. There was wood in abundance, dotted here and there with a profuse and careless luxuriance; there were rounded banks of hills, and meadows through which an ample stream came flowing with a queenly sweep, and with a sheen that caught every noontide, and repeated it in a glorious blazon of gold. It skirted the hills, it skirted the wood, and came with a gushing fulness upon the very margin of the quiet little houseyard that compassed the dwelling. And from the door, underneath cherry trees, one could catch glimpses of the great stretch of the Hudson into which the brook passed; and the farther shores were so distant, that the Hudson looked like a bay of the sea. A gaunt American who was in charge of the premises did the honors of the place, and in the intervals of expressing the juices from a huge quid of tobacco that lay in his cheek, he enlarged upon the qualities of the soil.

To him the view or situation was nothing, but the capacity for corn or rye was the main "p'int."

"Ef yer want a farm, Mister, yer want sile; now this 'ere (turning up a turf with a back thrust of his heel) is what I call sile; none o' yer dum leachy stuff; you put manure into this 'ere, and it stays 'put.'"

"Grows good crops, then," I threw in, by way of interlude.

"I guess it dooz, Mister. Corn, potatoes, garden sass,—why, only look at this 'ere turf; see them clovers, and this blue grass. Ef you was a farmer—doan't know but you be, but doan't look jist like one—you'd know that 'tain't every farm can scare up such a turf as that."

"Very true," I remark; while my lank friend adjusts his quid for a new bit of comment.

"Now here's Simmons on the hill—smart man enough, but doan't know nothing 'bout farmin'—them hills he's bought doan't bear nothin' but pennyrial; ten acres on't wouldn't keep a good cosset sheep." And my friend expectorates with a good deal of emphasis.

I suggested that many came into the country for good views and a fine situation.

"I know it, sir," said my lank friend; "this's a free country, and a man can do as he likes, leastwise we used to think so; but as for me, give me a good black sile 'bout seven inches thick, and good turf top on't, and a good smart team, and I take out my views, along in the fall o' the year, in the corn crib. Them's my sentiments."

I think I won upon my tall friend by expressing my approval of so sound opinions; and in the course of talk, we found ourselves again upon the dainty lawn by the doorstep, near to which the brook surged along, brimful and deep, to the river. Over-deep, indeed, it seemed, for so near neighborhood to the house. An expression of mine to this effect was amply confirmed by the tall farmer. Only a year or so gone, a little child had tumbled in, and was "drownded."

And this was perhaps the reason why the family left so attractive a place, I suggested.

"Oh Lord, no, sir; 'twas a pesky little thing, belonged down to the landin'. Fever-'nager's what driv the folks off, in my opinion."

"Ah, they do have the fever about here, then?"

"Gosh—Smithers here—p'raps you doan't know Smithers—no; waal, he's got it, got it bad, that's so; and what's wus, his chil'en s'got it, and his wife s'had it; and my wife here, a spell ago, what does she do, but up and takes it, s'bad s'enny on 'em; 'ts a dum curi's keind o' thing. You doan't know nothin' when 'ts comin'; and you doan't know no more when 'ts goin'; and arter 'ts dun, 'tain't no small shakes of a thing; a feller keeps keinder ailin'."

Upon a sudden the place took on a new aspect for me; its cool shade seemed the murky parent of miasma; the wind sighed through the leaves with a sickly sound, and the brook, that gave out a little while before a roistering cheerfulness in its dash, now surged along with only a quick succession of sullen plashes.

I must recur to one other disappointment in respect of a country place, which possessed every one of the features I had desired in unmistakable type; and yet all these so curiously distraught that they possessed no harmony or charm. I ought perhaps to except the sea view, which was wide to a fault, and so near that on turbulent days of storm, it must have created the illusion that you were fairly afloat.

A sight of the sea, to temper a fair landscape, and lend it ravishing reach to a far-off line of glistening horizon, is a very different thing from that bold, broadside, every-day nearness, which outroars all the pleasant land sounds, making your country quietude a mere fiction, and the broad presence of ocean the engrossing reality. So it was with the place of which I speak; beside this, the slope was slight and gradual—only one billowy lift—as if the land had some time caught the undulations of the sea after some heavy ground swell, and kept the uplift after the sea had settled to its fair-weather proportions. The brook was of an unnoticeable flow, that idled from a neighbor's grounds, and the wood, such as it was, only a spur of silver poplars that had stolen through from the same neighbor's territory, and had shot up into a white and tangled wilderness.

The occupant and owner of the place—of may be seventy acres—was one of those wiry, energetic, restless young men of New England stock, thrifty, shrewd, spurning all courtesies, bound to push on in life; a type of that nervous unrest by which God has peopled the West and California. Never gaining, but always despising, the calm that comes of satisfied endeavor, whether in the establishment of a home, or the accumulation of money, these fast ones are very confident in their ability withal, and in their judgment; making light of difficulties, full of contempt for all knowledge which has not shown practical and palpable conquests. The owner had planted his farm to vegetables—not an acre of it but bristled with some marketable crop; nearness to the city had warranted it, and "there was money in the business." To talk with such a man about comparative views, or situations, would have been to talk French with him. An unknown advertiser had demanded the very features embraced in his farm; there they were—sea enough, brook, wood, and slope. If I wished them enough to pay his price, I could have them. He felt quite sure that I should find nothing that came nearer the mark, and he argued the matter with a strenuous, earnest vehemence, that fairly enchained my attention; and while my admiring aspect seemed to yield assent to every presentation he made of the subject, and while, as in the case of the red-bearded German, there was a sort of magnetism that bound me to outer acquiescence, at the same time all my inner feeling was kindled into open revolt against the man's presumption, and his turnips, and his lines of cabbages, and his poplars, and near breadth of sea.

He did not sell to me; but I have no doubt that he sold; I have no doubt that he made money by his turnips, and more money by the sale of his land; and it would not surprise me to see him some day, if I go in that direction, speaker of the house of representatives in the State of Iowa, or Minnesota. There are men who carry in their presuming, restless energy the brand of success—not always an enviable one, still less frequently a moral one, but always palpable and noisy. Such a man makes capital fight with dangeı of all sorts ; he knows no yielding to fatigues—to any natural obstacles, or to conscience. It is hard to conceive of him as dying, without a sharp and nervous protest, which seems conclusive to his own judgment, against the absurd dispensations of Providence. Who does not see faces every day, whose eager, impassioned unrest is utterly irreconcilable with the calm long sleep we must all fall to at last?

But this story of unsuccessful experiences grows wearisome to me, and, I doubt not, to the reader. One after another the hopes I had built upon my hatful of responses, failed me. June was bursting every day into fuller and more tempting leafiness. The stifling corridors of city hotels, the mouldy smell of country taverns, the dependence upon testy Jehus, who plundered and piloted me through all manner of out-of-the-way places, became fatiguing beyond measure.

And it was precisely at this stage of my inquiry, that I happened accidentally to be passing a day at the Tontine inn, of the charming city of N—h—. (I use initials only, in way of respectful courtesy for the home of my adoption.) The old drowsy quietude of the place which I had known in other days, still lingered upon the broad green, while the mimic din of trade rattled down the tidy streets, or gave tongue in the shrill whistle of an engine. The college still seemed dreaming out its classic beatitudes, and the staring rectangularity of its enclosures and buildings and paths seemed to me only a proper expression of its old geometric and educational traditions.

Most people know this town of which I speak, only as a scudding whirl of white houses, succeeded by a foul sluiceway, that runs along the reeking backs of shops, and ends presently in gloom. A stranger might consider it the darkness of a tunnel, if he did not perceive that the railway train had stopped; and presently catch faint images of a sooty stairway, begrimed with smoke—up and down which dim figures pass to and fro, and from the foot of which, and the side of which, and all around which, a score of belching voices break out in a passionate chorus of shouts; as the eye gains upon the sootiness and gloom, it makes out the wispy, wavy lines of a few whips moving back and forth amid the uproar of voices; it lights presently upon the star of a policeman, who seems altogether in his element in the midst of the hurly-burly. Becloaked and shawled figures enter and pass through the carriages; they may be black, or white, or gray, or kinsfolk—you see nothing but becloaked figures passing through; portmanteaus fall with a slump, and huge dressing cases fall with a slam, upon what seems, by the ear, to be pavement; luggage trucks keep up an uneasy rattle; brakemen somewhere in still lower depths strike dinning blows upon the wheels, to test their soundness; newsboys, moving about the murky shades like piebald imps, lend a shrill treble to the uproar; the policeman's star twinkles somewhere in the foreground; upon the begrimed stairway, figures flit mysteriously up and down; there is the shriek of a steam whistle somewhere in the front; a shock to the train; a new deluge of smoke rolls back and around newsboys, police, cabmen, stairway, and all; there is a crazy shout of some official, a jerk, a dash—figures still flitting up and down the sooty stairway—and so, a progress into day (which seemed never more welcome). Again the backs of shops, of houses, heaps of debris, as if all the shop people and all the dwellers in all the houses were fed only on lobsters and other shellfish; a widening of the sluice, a gradual recovery of position to the surface of the ground—in time to see a few tall chimneys, a great hulk of rock, with something glistening on its summit, a turbid river bordered with sedges, a clump of coquettish pine trees—and the conductor tells you all this is the beautiful city of N—h—.

The natural impression of a stranger would be that the city was situated upon a considerable eminence, which had required deep boring for the proper adjustment of levels. The impression would be an unjust one; in all that dreary sink of a station, there is no height involved except the height of corporate niggardliness.* The town is as level as Runnimede.

A friend called upon me shortly after my arrival, and learning the errand upon which I had been scouring no inconsiderable tract of country, proposed to me to linger a day more, and take a drive about the suburbs. I willingly complied with his invitation; though I must confess that my idea of the suburbs, colored as it was by old recollections of college walks over dead stretches of level, in order to find some quiet copse, where I might bandy screams with a bluejay, in rehearsal of some college theme—all this, I say, moderated my expectations.

It seems but yesterday that I drove from among the tasteful houses of the town, which since my boy

* The mere fact that the roadbed is beneath the general level, is a source of great convenience, but affords no reason for converting the station into a reeking cellar. The purchase of certain adjoining premises, and the transfer to them of offices and refreshment rooms, and the glazing of the roof, would enable the company to pour down a flood of light upon the dreary depot, and give it a height and breadth and airiness, which so rich a corporation fairly owes to the comfort of the public, as well as the reputation of the city.

time had crept far out upon the margin of the plain. It seems to me that I can recall the note of an oriole, that sang gushingly from the limbs of an overreaching elm as we passed. I know I remember the stately broad road we took, and its smooth, firm macadam. I have a fancy that I compared it in my own mind, and not unfavorably, with the metal of a road, which I had driven over only two months before in the environs of Liverpool. I remember a somewhat stately country house that we passed, whose architecture dissolved any illusions I might have been under, in regard to my whereabouts. I remember turning slightly, perhaps to the right, and threading the ways of a neat little manufacturing village,—catching views of waterfalls, of tall chimneys, of open pasture grounds; and remember bridges, and other bridges, and how the village straggled on with its neat white palings, and whiter houses, with honeysuckles at the doors; and how we skirted a pond, where the pads of lilies lay all idly afloat; and how a great hulk of rock loomed up suddenly near a thousand feet, with dwarfed cedars and oaks tufting its crevices—tufting its top, and how we drove almost beneath it, so that I seemed to be in Meyringen again, and to hear the dash of the foaming Reichenbach; and how we ascended again, drifting through another limb of the village, where the little churches stood; and how we sped on past neat white houses,—rising gently,—skirted by hedgerows of tangled cedars, and presently stopped before a grayish-white farmhouse, where the air was all aflow with the perfume of great purple spikes of lilacs. And thence, though we had risen so little I had scarce noticed a hill, we saw all the spires of the city we had left, two miles away as a bird flies, and they seemed to stand cushioned on a broad bower of leaves; and to the right of them, where they straggled and faded, there came to the eye a white burst of water which was an arm of the sea; beyond the harbor and town was a purple hazy range of hills,—in the foreground a little declivity, and then a wide plateau of level land, green and lusty, with all the wealth of June sunshine. I had excuse to be fastidious in the matter of landscape, for within three months I had driven on Richmond hill, and had luxuriated in the valley scene from the *côte* of St. Cloud. But neither one or the other forbade my open and outspoken admiration of the view before me.

I have a recollection of making my way through the hedging lilacs, and ringing with nervous haste at the door bell; and as I turned, the view from the step seemed to me even wider and more enchanting than from the carriage. I have a fancy that a middle-aged man, with iron-gray whiskers, answered my summons in his shirt sleeves, and proposed joining me directly under some trees which stood a little way to the north. I recollect dimly a little country coquetry of his, about unwillingness to sell, or to name a price; and yet how he kindly pointed out to me the farm-lands, which lay below upon the flat, and the valley where his cows were feeding just southward, and how the hills rolled up grandly westward, and were hemmed in to the north by a heavy belt of timber.

I think we are all hypocrites at a bargain. I suspect I threw out casual objections to the house, and the distance, and the roughness; and yet have an uneasy recollection of thanking my friend for having brought to my notice the most charming spot I had yet seen, and one which met my wish in nearly every particular.

It seems to me that the ride to town must have been very short, and my dinner a hasty one: I know I have a clear recollection of wandering over those hills, and that plateau of farmland, afoot, that very afternoon. I remember tramping through the wood, and testing the turf after the manner of my lank friend upon the Hudson. I can recall distinctly the aspect of house, and hills, as they came into view on my second drive from the town; how a great stretch of forest, which lay in common, flanked the whole, so that the farm could be best and most intelligibly described as—lying on the edge of the wood; and it seemed to me, that if it should be mine, it should wear the name of Edgewood.

It is the name it bears now. I will not detail the means by which the coyness of my iron-gray-haired friend was won over to a sale; it is enough to tell that within six weeks from the day on which I had first sighted the view, and brushed through the lilac hedge at the door, the place, from having been the home of another, had became a home of mine, and a new stock of *Lares* was blooming in the *Atrium*.

In the disposition of the landscape, and in the breadth of the land, there was all, and more than I had desired. There was an eastern slope where the orchard lay, which took the first burst of the morning, and the first warmth of Spring; there was another valley slope southward from the door, which took the warmth of the morning, and which keeps the sun till night. There was a wood, in which now the little ones gather anemones in spring, and in autumn, heaping baskets of nuts. There was a strip of sea in sight, on which I can trace the white sails, as they come and go, without leaving my library chair; and each night I see the flame of a lighthouse kindled, and its reflection dimpled on the water. If the brook is out of sight, beyond the hills, it has its representative in the fountain that is gurgling and plashing at my door.

And it is in full sight of that sea, where even now the smoky banner of a steamer trails along the sky, and in the hearing of the dash of that very fountain, and with the fragrance of those lilacs around me, that I close this initial chapter of my book, and lay down my pen.

TAKING REINS IN HAND.

II.

TAKING REINS IN HAND.

Around the House.

ALTHOUGH possessing all the special requisites of which I had been in search, yet the farm was by no means without its inaptitudes and roughnesses. There was an accumulation of half-decayed logs in one quarter, of mouldering chips in another,—being monumental of the choppings and hewings of half a score of years. Old iron had its establishment in this spot; cast-away carts and sleds in that; walls which had bulged out with the upheaval of—I know not how many—frosts, had been ingeniously mended with discarded harrows or axles; there was the usual debris of clam shells, and there were old outbuildings standing awry, and showing rhomboidal angles in their outline. These approached the house very nearly,—so nearly, in fact, that in one direction at least, it was difficult to say where the province of the poultry and calves ended, and where the human occupancy began.

There was a monstrous growth of dock and burdock about the outer doors, and not a few rank shoots of that valuable medicinal herb—stramonium. There were the invariable clumps of purple lilacs, in most unmanageable positions; a few straggling bunches of daffodils; an ancient garden with its measly looking, mossy gooseberries; a few strawberry plants, and currant bushes keeping up interruptedly the pleasant formality of having once been set in rows, and of having nodded their crimson tassels at each other across the walk. There were some half dozen huge old pear trees, immediately in the rear of the house, mossy, and promising inferior native fruit; but full of a vigor that I have since had the pleasure of transmuting into golden Bartletts. There were a few plum trees, loaded with black knot; a score of peach trees in out of the way places, all showing unfortunate marks of that vegetable jaundice, the yellows, which throughout New England is the bane of this delicious fruit.

There was the usual huge barn, a little wavy in its ridge, and with an aged settle to its big doors; while under the eaves were jagged pigeon holes, cut by adventurous boys, ignorant of curvilinear harmonies. Upon the peak was a lively weather-cock of shingle, most preposterously active in its motions, and trimming to every flaw of wind with a nervous rapidity, that reminded me of nothing so much as of the alacrity of a small newspaper editor. There was the attendant company of farm sheds, low sheds, high sheds, tumble-down sheds, one with a motley array of seasoned lumber, well dappled over with such domestic coloring as barn-yard fowls are in the habit of administering; another, with sleds and sleighs,—looking out of place in June—and submitted to the same domestic garniture. There was the cider mill with its old casks, and press, seamy and mildewed, both having musty taint. A convenient mossy cherry tree was hung over with last year's scythes and bush-hooks, while two or three broken ox chains trailed from the stump of a limb, which had suffered amputation. Nor must I forget the shop, half home-made, half remnant of something better, with an old hat or two thrust into the broken sashes—with its unhelved, gone-by axes, its hoes with half their blade gone, its dozen of infirm rakes, its hospital shelf for broken swivels, heel-wedges and dried balls of putty.

I remember passing a discriminating eye over the tools, bethinking me how I would swing the broad-axe, or put the saws to sharp service; for in bargain ing for the farm, I had also bargained for the implements of which there might be immediate need.

Directly upon the roadway, before the house, rose a high wall, supporting the little terrace that formed the front yard; the terrace was a wilderness of roses, lilacs, and unclipped box. The entrance way was by a flight of stone steps which led through the middle of the terrace, and of the wall; while over the steps hung the remnants of an ancient archway, which had once supported a gilded lantern; and I was told with an air of due reverence, that this gilded spangle of the town life, was a memento of the hospitalities of a certain warm-blooded West Indian, who in gone by years had lighted up the country home with cheery festivities. I would have cherished the lantern if it had not long before disappeared; and the steps that may have once thronged under it, must be all of them heavy with years now, if they have not rested from their weary beat altogether. Both wall and terrace are now gone, and a gentle swell of green turf is in their place, skirted by a hedge and low rustic paling, and crowned by a gaunt pine tree, and a bowering elm.

The same hospitable occupant, to whom I have referred, had made additions to the home itself, so as to divest it of the usual, stereotyped farm-house look, by a certain quaintness of outline. This he had done by extending the area of the lower story some ten feet, in both front and rear, while the roof of this annex was concealed by a heavy balustrade, perched upon the eaves; thus giving the effect of one large cube, surmounted by a lesser one; the uppermost was topped with a roof of sharp pitch, through whose ridge protruded two enormous chimney stacks. But this alteration was of so old a date as not to detract from the venerable air of the house. Even the jaunty porch which jutted in front of all, showed gaping seams, and stains of ancient leakage, that forbade any suspicion of newness.

Within, the rooms had that low-browed look which belongs to country farm-houses; and I will not disguise the matter by pretending that they are any higher now. I have occasional visitors whom I find it necessary to caution as they pass under the doorways; and the stray wasps that *will* float into the open casements of so old a country house, in the first warm days of Spring, are not out of reach of my boy, (just turned of five,) as he mounts a chair, and makes a cut at them with his dog-whip, upon the ceiling.

I must confess that I do not dislike this old humility of house-building; if windows, open chimney places, and situation give good air, what matters it, that your quarters by night are three or four feet nearer to your quarters by day? In summer, if some simple trellised pattern of paper cover the ceiling, you enjoy the illusion of a low branching bower; and of a winter evening, the play of the fire-light on the hearth flashes over it, with a kindly nearness.

I know the outgoing parties found no pleasant task in the leave-taking. I am sure the old lady who was its mistress felt a pang that was but poorly concealed; I have a recollection that on one of my furtive visits of observation, I unwittingly came upon her—at a stand-still over some bit of furniture that was to be prepared for the cart,—with her handkerchief fast to her eyes. It cannot be otherwise at parting with even the lowliest homes, where we have known of deaths, and births, and pleasures, and little storms that have had their sweep and lull; and where slow-pacing age has declared itself in gray hair, and the bent figure. It is tearing leaf on leaf out of the thin book where our lives are written.

Even the farmer's dog slipped around the angles of the house, as the change was going forward, with a fitful, frequent, uneasy trot, as if he were disposed to make the most of the last privileges of his home. The cat alone, of all the living occupants, took matters composedly, and paced eagerly about from one to another of her disturbed haunts in buttery and kitchen, with a philosophic indifference. I should not wonder indeed if she indulged in a little riotous exultation at finding access to nooks which had been hitherto cumbered with assemblages of firkins and casks. I have no faith in cats: they are a cold-blooded race; they are the politicians among domestic animals; they care little who is master, or what are the over-turnings, so their pickings are secure; and what are their midnight caucuses but primary meetings?

My Bees.

A SHELF, on which rested five bee-hives with their buzzing swarms, stood beside a clump of lilacs, not far from one of the side doors of the farm house. These the outgoing occupant was indisposed to sell; it was "unlucky," he said, to give up ownership of an old-established colony. The idea was new to me, and I was doubly anxious to buy, that I might give his whimsey a fair test. So I overruled his scruples at length, moved the bees only a distance of a few yards, gave them a warm shelter of thatch, and strange to say, they all died within a year.

I restocked the thatched house several times afterward; and there was plenty of marjoram and sweet clover to delight them; whether it was that the misfortunes of the first colony haunted the place, I know not, but they did not thrive. Sometimes, I was told, it was the moth that found its way into their hives; sometimes it was an invasion of piratical ants; and every summer I observe that a few gallant king birds take up their station near by, and pounce upon the flying scouts, as they go back with their golden booty.

I have not the heart to shoot the king birds; nor do I enter very actively into the battle of the bees against the moths, or the ants; least of all, do I interfere in the wars of the bees among themselves, which I have found, after some observation, to be more destructive and ruinous, than any war with foreign foes.* I give them fair play, good lodging, limitless flowers, willows bending (as Virgil advises) into the quiet water of a near pool; I have even read up the stories of poor blind Huber, who so loved the bees, and the poem of Giovanni Rucellai, for their benefit: if they cannot hold their sceptre against the tender-winged moths, who have no cruel stings, or against the ants, or the wasps, or give over their satanic quarrels with their kindred, let them abide the consequences. I will not say, however, but that

* The Rev. Charles Butler, in his "Feminine Monarchie" (London, 1609), after speaking in Chapter VII of "Deir Enemies," continues: "But not any one of des[e], nor all des[e] togeder, doo half so muc harm to de Bees, as de Bees. *Apis api, ut homo homini, Lupus.* Dey mak de greatest spoil bot of bees and of hoonie. Dis robbing is practised all de yeer."

the recollection of the sharp screams of a little "curl pate" that have once or twice pierced my ears, as she ventured into too close companionship, has indisposed me to any strong advocacy of the bees.

My experience enables me to say that hives should not be placed too near each other; the bees have a very human propensity to quarrel, and their quarrels are ruinous. They blunder into each other's homes, if near together, with a most wanton affectation of forgetfulness; and they steal honey that has been carefully stored away in the cells of sister swarms, with a vicious energy that they rarely bestow upon a flower. In their field forays, I believe they are respectful of each other's rights; but at home, if only the order is once disturbed, and a neighbor swarm shows signs of weakness, they are the most malignant pirates it is possible to conceive of.

Again, let no one hope for success in their treatment, unless he is disposed to cultivate familiarity; a successful bee-keeper loves his bees, and has a way of fondling them, and pushing his intimacy about the swarming time, which I would not counsel an inapt or a nervous person to imitate.

Gélieu, a Swiss authority, and a rival of Huber in his enthusiasm, says: "Beaucoup de gens aiment les abeilles; je n'ai vu personne qui les aima *médiocrement;* on se *passionne* pour elles."

I have a neighbor, a quiet old gentleman, who is possessed of this passion; his swarms multiply indefinitely; I see him holding frequent conversations with them through the backs of their hives; all the stores of my little colony would be absorbed in a day, if they were brought into contact with his lusty swarms.

Many of the old writers tell pleasant stories of the amiable submission of their favorites to gentle handling; but I have never had the curiosity to put this submission to the test. I remember that Van Amburgh tells tender stories of the tigers.

I have observed, however, that little people listen with an amused interest to those tales of the bees, and I have sometimes availed myself of a curious bit of old narrative, to staunch the pain of a sting.

"Who will listen," I say, "to a story of M. Lombard's, about a little girl, on whose hand a whole swarm of bees once alighted?"

And all say "I"—save the sobbing one, who looks consent.

M. Lombard was a French lawyer, who was for a long time imprisoned in the dungeons of Robespierre; and when that tyrant reformer was beheaded, this prisoner gained his liberty, and went into the country, where he became a farmer, and wrote three or four books about the bees: among other things he says:

"A young girl of my acquaintance was greatly afraid of bees, but was completely cured of her fear by the following incident. A swarm having left a hive, I observed the queen alight by herself, at a little distance from the apiary. I immediately called my little friend, that I might show her this important personage; she was anxious to have a nearer view of her majesty, and therefore, having first caused her to draw on her gloves, I gave the queen into her hand. Scarcely had I done so, when we were surrounded by the whole bees of the swarm. In this emergency, I encouraged the trembling girl to be steady, and to fear nothing, remaining myself close by her, and covering her head and shoulders with a thin handkerchief. I then made her stretch out the hand that held the queen, and the bees instantly alighted on it, and hung from her fingers as from the branch of a tree. The little girl, experiencing no injury, was delighted above measure at the novel sight, and so entirely freed from all fear, that she bade me uncover her face. The spectators were charmed at the interesting spectacle. I at length brought a hive, and shaking the swarm from the child's hand, it was lodged in safety without inflicting a single sting."

As I begin the story, there is a tear in the eye of the sobbing one, but as I read on, the tear is gone, and the eye dilates; and when I have done, the sting is forgotten.

I have written thus at length, at the suggestion of my thatch of a bee house, because I shall have nothing to say of my bees again, as co-partners with me in the flowers, and in the farm. I have to charge to their account a snug sum for purchase money, and for their straw housing—a good many hours of bad humor, and the recollection of those little screams to which I have already alluded. Thus far, I can only credit them with one or two moderately sized jars of honey, and a pleasant concerted buzzing with which they welcome the first warm weather of the Spring. Even as I write, I observe that a few of my winged workers are alight upon the mossy stones that lie half covered in the basin of the fountain, and are sedulously exploring the water.

Clearing Up.

OF course one of the first aims, in taking possession of such a homestead as I have partially described, was to make a clearance of debris, of unnecessary palings, of luxuriant corner crops of nettles and burdocks, of mouldering masses of decayed vegetable matter, of old conchologic deposits, and ferruginous wreck; all this clearance being not so much agricultural employment, as hygienic. There seems to have been a mania with the old New England householders, in the country, for multiplying enclosures,—front yards, back yards, south and north yards,—all with their palings and gates, which grow shaky with years, and give cover to rank and worthless vegetation in corners, that no cultivation can reach. Of this multitude of palings I made short work: good taste, economy, and all rules of good tillage, unite in favor of the fewest possible enclosures, and confirm the wisdom of making the palings for such as are necessary, as simple as their office of defence will allow.

So it happened under my ruling that the little terrace yard of the front lost its identity, and was merged in the yard to the north,—with the little bewildered garden to the south,—with the straggling peach orchard in the rear; and all these merged again, by the removal of a tottling wall, with the valley pasture that lay southward; where now clumps of evergreen, and azalias, and lilacs crown the little swells, and hide the obtrusive angles of barriers beyond; so that the children may race, from the door, over firm, clean, green sward, for a gunshot away. This change has not been only to the credit of the eye, but in every particular economic. The cost of establishing and repairing the division palings has been done away with; the inaccessible angles of enclosures which fed monstrous wild growth, are submitted to even culture and cropping; an under drain through the bottom of the valley lawn, has absorbed the scattered stones and the tottling wall of the pasture, and given a rank growth of red-top and white clover, where before, through three months of the year, was almost a quagmire. This drain, fed by lesser branches laid on from time to time through the springy ground of the peach orchard, and by the waste way of the fountain at the door, now discharges into a little pool (once a mud hole) at the extremity of the lawn, where a willow or two timidly dip their branches, and the frogs welcome every opening April with a riotous uproar of voices. Even the scattered clumps of trees stand upon declivities where cultivation would have been difficult, or they hide out-cropping rocks which were too heavy for the walls, or the drains. So it has come about that the old flimsy pasture, with its blotches of mulleins, thistles, wax myrtles, and the ill shapen yard, straggling peach orchard (long since gone by), have made my best grass field, which needs only an occasional top dressing of ashes or compost, and a biennial scratching with a fine-toothed harrow, to yield me two tons to the acre of sweet-scented hay.

I may remark here, in way of warning to those who undertake the renovation of slatternly country places, with exuberant spirits, that it is a task which often seems easier than it proves. More especially is this the case where there is an accumulation of old walls, and of unsightly, clumsy-shaped rocks to be dealt with. They may indeed be transferred to new walls; but this involves an expenditure, oftentimes, which no legitimate estimate of a farm revenue will warrant; and I propose to illustrate in this book no theories of improvement, whether as regards ornamentation or increased productiveness, which a sound economy will not authorize. Agricultural successes which are the result of simple, lavish expenditure, without reference to agricultural returns, are but empty triumphs; no success in any method of culture is thoroughly sound and praiseworthy, except it be imitable, to the extent of his means, by the smallest farmer. The crop that is grown at twice its market value to the bushel, may possibly suggest a hint to the scientific theorist; but it will never be emulated by the man whose livelihood depends upon the product of his farm. Those who transfer the accumulated fortunes of the city to the country, for the encouragement of agriculture, should bear in mind, first of all, that their endeavors will have healthy influence, only so far as they are imitable;

and they will be imitable only so far as they are subordinated to the trade laws of profit and loss. Farming is not a fanciful pursuit; its aim is not to produce the largest possible crop at whatever cost; but its aim is, or should be, taking a series of years together, to produce the largest crops at the *least* possible cost.

If my neighbor, by an expenditure of three or four hundred dollars to the acre in the removal of rocks and other *impedimenta*, renders his field equal to an adjoining smooth one, which will pay a fair farm rental on a valuation of only two hundred dollars per acre, he may be congratulated upon having extended his available agricultural area, but he cannot surely be congratulated on having made a profitable transaction.

The weazen faced old gentlemen who drive by in their shirt sleeves, and call attention to the matter with a gracious wave of their hickory whipstocks, allow that—"it's fine; but don't pay." Such observers—and they make up the bulk of those who have the country in their keeping—must be addressed through their notions of economy, or they will not be reached at all.

In the case supposed, I have, of course, assumed that only ordinary farm culture was to be bestowed: although there may be conditions of high tillage, extraordinary nicety of culture, and nearness to a large market, which would warrant the expenditure of even a thousand dollars per acre with profitable results.

But rocky farms, even away from markets, are not without their profits, and a certain wild, yet subdued order of their own. I have never seen sweeter or warmer pasture ground, than upon certain hillsides strown thick with great granite boulders, spangled with mica, and green-gray mosses; nor was the view unthrifty, with its fat, ruffle-necked merino ewes grazing in company; nor yet unattractive to other than farm-eyes—with its brook bursting from under some ledge that is overhung with gnarled birches, and illuminated with nodding, crimson columbines—then yawing away between its green banks, with a new song for every stone that tripped its flow.

One of the daintiest and most productive fruit gardens it was ever my pleasure to see, was in the midst of other gray rocks; the grape vines so trained as to receive the full reflection of the sun from the surface of the boulders, and the intervals occupied with rank growing gooseberries and plums, all faithfully subject to spade culture. The expense of the removal of the rocks would have been enormous; and I doubt very seriously if the productive capacity would have been increased. Again, I have seen a ridge of cliff with its outlying slaty debris, in the very centre of a garden, which many a booby leveller would have been disposed to blast away, and transmute into walls,—yet under the hand of taste, so tressed over with delicate trailing plants, and so kindled up with flaming spikes of salvia, and masses of scarlet geranium, as to make it the crowning attraction of the place. All clearance is not judicious clearance.

But I have not yet cleared the way to my own back door; though at a distance of only a few rods from the highway, I could reach it, on taking occupancy, only by skirting a dangerous looking shed, and passing through two dropsical gates that were heavy with a mass of mouldy lumber.

These gates opened upon a straggling cattle yard, whose surface was so high and dense, as to distribute a powerful flow of yellow streamlets in very awkward directions after every shower. One angle of this yard it was necessary to traverse before reaching my door. My clearance here was decisive and prompt. The threatening shed came down upon the run; the mouldering gates and fences were splintered into kindling wood; the convexity of the cattle yard was scooped into a dish, with provision for possible overflow in safe directions. A snug compact fence blinded it all, and confined it within reasonable limits. A broad, free, gravelly yard, with occasional obtrusive stones, now lay open, through which I ordered a loaded team to be driven by the easiest track from the highway to the door, and thence to make an easy and natural turn, and pass on to the stable-court. This line of transit marked out my road: what was easiest for the cattle once, would be easiest always. There is no better rule for laying down an approach over rolling ground—none so simple; none which, in one instance out of six, will show more grace of outline. The obtrusive stones were removed; the elliptical spaces described by the inner line of track, which were untouched, and which would need never to be touched by any passage of teams, were dug over and stocked with evergreens, lilacs, and azalias.

These are now well established clumps, in which wild vines have intruded, and under which the brood of summer chickens find shelter from the sun, and the children a pretty cover for their hoydenish "hide and go seek."

Thus far I have anticipated those changes and improvements which immediately concerned the comfort and the order of the home. With these provided, and the paperers and painters all fairly turned adrift, and the newly planted flowers abloom, the question occurs—what shall be done with the Farm?

What to Do with the Farm.

THERE are not a few entertaining people of the cities, who imagine that a farm of one or two hundred acres has a way of managing itself; and that it works out crops and cattle from time to time, very much as small beer works into a foamy ripeness, by a law of its own necessity.

I wish with all my heart that it were true; but it is not. For successful farming, there must be a well digested plan of operations, and the faithful execution of that plan. It is possible, indeed, to secure the services of an intelligent manager, upon whom shall devolve all the details of the business, and who shall shape all the agricultural operations, by the rules of his own experience; but however extended this experience may have been, the result will be, in nine cases out of ten, most unsatisfactory to one who wishes to have a clear and intimate knowledge of the capabilities of his land; and very disagreeably unsatisfactory to one who has entertained the pleasing illusion that farm lands should not only be capable of paying their own way, but of making respectable return upon the capital invested. Your accomplished farm manager—usually of British birth and schooling, but of a later American finish,—is apt to entertain the conviction that an employer who gives over farm land to his control, regards such farm land only as a pleasant parade ground for fine cattle and luxuriant crops, which are to be placed on show without much regard to cost. And if he can establish the owner in a conspicuous position on the prize lists of the County or State Societies, and excite the gaping wonderment of old-fashioned neighbors by the luxuriance of his crops, he is led to believe that he has achieved the desired success.

The end of it is, that the owner enjoys the honors of official mention, without the fatigue of relieving himself of ignorance; the manager is doubly sure of his stipend; and the inordinate expense under a direction that is not limited by commercial proprieties or proportions, weakens the faith of all onlookers in "improved farming."

I am satisfied that a great deal of hindrance is done in this way to agricultural progress, by those who have only the best intentions in the matter. My friend, Mr. Tallweed, for instance, after accumulating a fortune in the city, is disposed to put on the dignity of country pursuits, and advance the interests of agriculture. He purchases a valuable place, builds his villa, plants, refits, exhausts architectural resources in his outbuildings, all under the advice of a shrewd Scotchman recommended by Thorburn, and can presently make such show of dainty cattle, and of mammoth vegetables, as excites the stare of the neighborhood, and leads to his enrolment among the dignitaries of the County Society.

But the neighbors who stare, have their occasional chat with the canny Scot, from whom they learn that the expenses of the business are "gay large;" they pass a quiet side wink from one to the other, as they look at the vaulted cellars, and the cumbrous machinery; they remark quietly that the multitude of implements does not forbid the employment of a multitude of farm "hands;" they shake their heads ominously at the extraordinary purchases of grain; they observe that the pet calves are usually indulged with a wet nurse, in the shape of some rawboned native cow, bought specially to add to the resources of the fine-blooded dam; and with these things in their mind—they reflect.

If the results are large, it seems to them that the means are still more extraordinary; if they wonder at the size of the crops, they wonder still more at the liberality of the expenditure; it seems to them, after full comparison of notes with the "braw" Scot, that even their own stinted crops would show a better balance sheet for the farm. It appears to them that if premium crops and straight-backed animals can only be had by such prodigious appliances of men and money, that fine farming is not a profession to grow rich by. And yet, our doubtful friends of the homespun will enjoy the neighborhood of such a farmer, and profit by it; they love to sell him "likely young colts;" they eagerly furnish him with butter (at the town price), and possibly with eggs; his own fowls being mostly fancy ones, bred for premiums, and indisposed to lay largely; in short, they like to tap his superfluities in a hundred ways. They admire Mr. Tallweed, particularly upon Fair days, when he appears in the dignity of manager for some special interest; and remark, among themselves, that "the Squire makes a thunderin' better committee-man, than he does farmer." And when they read of him in their agricultural journal—if they take one—as a progressive, and successful agriculturist, they laugh a little in their sleeves in a quiet way, and conceive, I am afraid, the same unfortunate distrust of the farm journal, which we all entertain—of the political ones.

Yet the Squire is as innocent of all deception, and of all ill intent in the matter, as he is of thrift in his farming. Whoever brings to so practical a business the ambition to astonish by the enormity of his crops, at whatever cost, is unwittingly doing discredit to those laws of economy, which alone justify and commend the craft to a thoroughly earnest worker.

Having brought no ambition of this sort to my

trial of country life, even if I had possessed the means to give it expression, I had also no desire to give over all plans of management to a bailiff, however shrewd. The greatest charm of a country life seems to me to spring from that familiarity with the land, and its capabilities, which can come only from minute personal observation, or the successive developments of one's own methods of culture. I can admire a stately crop wherever I see it; but if I have directed the planting, and myself applied the dressing, and am testing my own method of tillage, I look upon it with a far keener relish. Every week it unfolds a charm; if it puts on a lusty dark green, I see that it is taking hold upon the fertilizers; if it yellows in the cool nights, and grows pale, I bethink me if I will not put off the planting for a week in the season to come; if it curl overmuch in the heats of later June, I reckon up the depth of my ploughing; and when the spindles begin to peep out from their green sheaths day after day, and lift up, and finally from their feathery fingers shake down pollen upon the silk nestling coyly below, I see in it all a modest promise to me—repeated in every shower—of the golden ears that shall by and by stand blazing in the October sunshine.

But all this only answers negatively my question of—what to do with the Farm?

At least, it shall not be handed over absolutely to the control of a manager, no matter what good character he may bring; and I will aim at a system of cropping, which shall make some measurable return for the cost of production.

Dairying.

ANY judicious farm-system must be governed in a large degree by the character of the soil, and by the nearest available market. It is not easy to create a demand for what is not wanted; nor is it much easier so to transmute soils by culture or by dressings, as to produce profitably those crops to which the soils do not naturally incline. I am fully aware that in saying this, I shall start an angry buzz about my ears, of those progressive agriculturists, who allege that skilful tillage will enable a man to produce any crop he chooses: I am perfectly aware that Tull, who was the great farm reformer of his day, ridiculed with unction what he regarded as those antiquated notions of Virgil, that soils had their antipathies and their likings, and that a farmer could not profitably impress ground to carry a crop against its inclination. But I strongly suspect that Tull, like a great many earnest reformers, in his advocacy of the supreme benefit of tillage, shot beyond the mark, and assumed for his doctrine a universality of application, which practice will not warrant. I am perfectly confident that no light and friable soil will carry permanent pasture or meadow, with the same profit which belongs to the old grass bottoms of the Hartford meadows, of the blue-grass region, and of Somersetshire. I am equally confident that no stiff clayey soil will pay so well for the frequent workings which vegetable culture involves, as a light loam.

Travellers who are trustworthy, tell us that the grape from which the famous Constantia wine is made, at the Cape of Good Hope, is grown from the identical stock which, on the Rhine banks, makes an inferior and totally different wine: and my own observation has shown me that the grapes which on the Lafitte estate make that ruby vintage whose aroma alone is equal to a draught of ordinary Medoc—only across the highway, and within gunshot, produce a wine for which the proprietor would be glad to receive a fourth only of the Lafitte price.

Lands have their likings then, though Mr. Tull be of the contrary opinion. Any crop may indeed be grown wherever we supply the requisite conditions of warmth, moisture, depth of soil, and appropriate dressings; but just in the proportion that we find these conditions absent in any given soil, and are compelled to supply them artificially, we diminish the chances of profit.

My own soil was of a light loamy character, and the farm lay within two miles of a town of forty thousand inhabitants.

Such being the facts, what should be the general manner of treatment?

Grazing, which is in many respects the most inviting of all modes of farming, was out of the question, for the reason that the soil did not incline to that firm, close turf-surface, which invites grazing, and renders it profitable. Nor do I mean to admit, what many old-fashioned gentlemen are disposed to affirm, that all land which does not so incline, is necessarily inferior to that which does. If grazing were the chiefest of agricultural interests, it might be true. But it must be observed that strong grass lands have generally a tenacity and a retentiveness of moisture, which forbid that frequent and early tillage, that is essential to other growths; and upon careful reckoning, I doubt very much, if it would not appear that some of the very light lands in the neighborhood of cities, pay a larger percentage upon the agricultural capital invested, than any purely grazing lands in the country. Again, even supposing that the soil were adapted to grazing, it is quite doubtful if the best of grazing lands will prove profitable in the neighborhood of large towns; doubtful if beef and mutton cannot be made cheaper in out-of-the-way districts, where by reason of distance from an every-day market, lands command a low price.

For kindred reasons, no farm, so near a large town of the East, invites the growth of grain: on this score there can be no competition with the West, except in retired parts of the country, where land is of little marketable value.

What then? Grazing does not promise well; nor does grain-growing. Shall I stock my land with grass, and sell the hay? Unfortunately, this experiment has been carried too far already. A near market, and the small amount of labor involved, always encourage it. But I am of opinion that no light land will warrant this strain, except where manures from outside sources are easily available, and are applied with a generous hand. Such, for instance, is the immediate neighborhood of the sea shore, where fish and rockweed are accessible; or, what amounts to the same thing, such disposition of the land as admits of thorough irrigation. In my case, both these were wanting. I must depend for manurial resources upon the consumption of the grasses at home.

And this suggests dairying: dairying in its ordinary sense, indeed, as implying butter and cheese making, involves grazing; and can be most profitably conducted on natural grass lands, and at a large distance from market, since the transport of these commodities is easy. But there remains another branch of dairying—milk supply—which demands nearness to market, which is even more profitable, and which does not involve necessarily a large reach of grazing land; the most successful milk dairies in this country, as in Great Britain, being now conducted upon the soiling principle—that is, the supply of green food to the cows, in their enclosures or stalls.

What plan then could be better than this? Transportation to market was small; the demand constant; the thorough tillage which the condition of the soil required, was encouraged; an accumulation of fertilizing material secured.

The near vicinity of a town suggests also to a good husbandman, the growth of those perishable products which will not bear distant transportation, such as the summer fruits and vegetables. These demand also a thorough system of tillage, and a light friable soil is, of all others, best adapted to their successful culture. But on the other hand, they do not in themselves furnish the means of recuperating lands which have suffered from injudicious overcropping. Their cultivation, unless upon fields which are already in a high state of tilth, involves a large outlay for fertilizing materials and for labor—which at certain seasons must be at absolute command.

In view of these considerations, which I commend to the attention and to the criticism of the Agricultural Journals, I determined that I would have my herd of milch cows, and commence professional life as milkman; keeping, however, the small fruits and the vegetables in reserve, against the time when the land by an effective recuperative system, should be able to produce whatever the market might demand.

Happily, too, a country liver is not bound to a single farm adventure. If the cows stand sweltering in the reeking stables, it shall not forbid a combing down of the ancient pear trees, and the tufting of all their tops with an abounding growth of new wood, that shall presently be aglow with the Bonne de Jersey, or with luscious Bartletts.

If there is a rattle of tins in the dairy, the bluebirds are singing in the maples. If an uneasy milker kicks over the pail, there is a patch of Jenny Linds that make a fragrant recompense. If the thunder sours the milk, the nodding flowers and the rejoicing grass give the shower a welcome.

Laborers

HAVING decided upon a plan, the next thing to be considered is the personal agency for its administration.

There was once a time, if we may believe a great many tender pastorals and madrigals such as Kit Marlowe sang, when there were milkmaids: and the sweetest of Overbury's "Characters" is his little sketch of the 'faire damsel,' who hath such fingers "that in milking a cow, it seemes that so sweet a milk-presse makes the milk the whiter or sweeter." But milkmaids now-a-days are mostly Connaught men, in cowhide boots and black satin waistcoats, who say "begorra," and beat the cows with the milking stool.

Overbury says of the ancient British type—"Her breath is her own, which sents all the yeare long of June, like a newmade haycock."

And I may say of the present representative—His breath is his own, which 'sents all the yeare long' of proof spirits, like a newmade still.

Overbury tenderly says—"Thus lives she, and all her care is she may die in the spring time, to have store of flowers stucke upon her winding sheet."

And I, as pathetically:—Thus fares he, and all

his care is he may get his full wage, and a good jollification 'nixt St. Parthrick's day.'

This is only my way of introducing the labor question, which, in every aspect, is a serious one to a party entering upon the management of country property. If such party is anticipating the employment of one of Sir Thomas Overbury's milk-maydes, or of the pretty damsel who sang Marlowe's song to Izaak Walton, let him disabuse his mind. In place of it all, he will sniff boots that remind of a damp cattle yard, and listen to sharp brogue that will be a souvenir of Donnybrook Fair. In briefest possible terms, the inferior but necessary labor of a farm must be performed now, in the majority of cases, by the most inefficient of Americans, or by the rawest and most uncouth of Irish or Germans.

There lived some twenty or thirty years ago in New England, a race of men, American born, and who, having gone through a two winters' course of district school ciphering and reading, with cropped tow heads, became the most indefatigable and ingenious of farm workers. Their hoeing was a sleight of hand; they could make an ox yoke, or an axe helve on rainy days; by adroit manipulation, they could relieve a choking cow, or as deftly, hive a swarm of bees. Their furrows indeed were not of the straightest, but their control of a long team of oxen was a miracle of guidance. They may have carried a bit of Cavendish twist in their waistcoat pockets; they certainly did not waste time at lavations; but as farm workers they had rare aptitude; no tool came amiss to them; they cradled; they churned, if need were; they chopped and piled their three cords of wood between sun and sun. With bare feet, and a keen-whetted six-pound Blanchard, they laid such clean and broad swaths through the fields of dewy herdsgrass, as made "old-country-men" stare. By a kind of intuition, they knew the locality of every tree, and of every medicinal herb that grew in the woods. Rarest of all which they possessed, was an acuteness of understanding, which enabled them to comprehend an order before it was half uttered, and to meet occasional and unforeseen difficulties, with a steady assurance, as if they had been an accepted part of the problem. It was possible to send such a man into a wood with his team, to select a stick of timber, of chestnut or oak, that should measure a given amount; he could be trusted to find such,—to cut it, to score it, to load it; if the gearing broke, he could be trusted to mend it; if the tree lodged, he could be trusted to devise some artifice for bringing it down; and finally, —for its sure and prompt delivery at the point indicated. Your Irishman, on the other hand, balks at the first turn; he must have a multitude of chains;

he needs a boy to aid him with the team, and another to carry a bar; he spends an hour in his doubtful estimate of dimensions; but "begorra, its a lumpish tree," and he thwacks into the rind a foot or two from the ground, so as to leave a 'nate' Irish stump. Half through the bole, he begins to doubt if it be indeed a chestnut or a poplar; and casting his eye aloft to measure it anew, an ancient woodpecker drops something smarting in his eye; and his howl starts the ruminating team into a confused entanglement among the young wood. Having eased his pain, and extricated his cattle, he pushes on with his axe, and presently, with a light crash of pliant boughs, his timber is lodged in the top of an adjoining tree. He tugs, and strains, and swears, and splits the helve of his axe in adapting it for a lever, and presently, near to noon, comes back for three or four hands to give him a boost with the tree. You return —to find the team strayed through a gate left open, into a thriving cornfield, and one of your pet tulip trees lodged in a lithe young hickory.

"Och! and it's a toolip—it is! and I was thinkin' 'twas niver a chistnut; begorra, it's lucky thin, it didn't come down intirely."

These and other such, replace the New-Englander born, who long ago was paid off, wrapped his savings in a dingy piece of sheepskin, scratched his head reflectingly, and disappeared from the stage. He has become the father of a race that is hewing its way in Oregon, or he is a dignitary in Wisconsin, or thwacking terribly among the foremost fighters of the war.

Here and there remains an aged representative of the class, with all his nasal twang and his aptitude for a score of different services; but the chances are, if he has failed of placing himself in the legislative chambers of the West, or of holding ownership of some rough farm of his own, that he has some moral obliquity which makes him an outcast.

Certain it is, that very few native Americans of activity and of energy are to be decoyed into the traces of farm labor, unless they can assume the full direction. American blood is fast, and fast blood is impatient with a hoe among small carrots. It is well, perhaps, that blood is so fast, and hopes so tall. These tell grandly in certain directions, but they are not available for working over a heap of compost. The American eagle is (or was) a fine bird, but he does not consume grasshoppers like a turkey.

In view of the fact that dexterous labor is not now available, there is a satisfaction in knowing that the necessity for it is year by year diminishing. Upon the old system of growing all that a man might need within his own grounds, a proper farm education embraced a considerable knowledge of a score of different crops and avocations. The tendency is now, however, to centralize attention upon that line of cropping which is best suited to the land; this limits the range of labor, while the improved mechanical appliances fill a thousand wants, which were once only to be met by a dexterous handicraft at home. None but a few weazen-faced old gentlemen of a very ancient school, think now-a-days of making their own ox yokes or their own cheese presses; or, if their crop be large, of pounding out their grain with a flail. And it is noticeable in this connection, that the implements in the use of which the native workers were most unmatchable, are precisely the ones which in practical farming are growing less and less important every year; to wit, the axe and the scythe: the first being now confined mostly to clearings of timber, and the second is fast becoming merely a garden implement for the dressing of lawns.

I perceive, very clearly, from all this, that I am not to be brought in contact with a race of Arcadians. Melibœus will not do the milking, nor Tityrus,—though there shall be plenty of snoozing under the beech trees. It is also lamentably true that the uncouth and unkempt Irish or Germans, whom it becomes necessary to employ, place no pride or love in their calling like the English farm laborers, or like that gone-by stock of New England farm workers at whom I have hinted.

Your Irish friend may be a good reaper, he may possibly be a respectable ploughman (though it is quite doubtful); but in no event will he cherish any engrossing attachment to country labors; nor will he come to have any pride in the successes that may grow out of them.

Every month he is ready to drift away toward any employment which will bring increase of pay. He is your factotum to-day, and to-morrow may be shouldering a hod, or scraping hides for a soap boiler. The German, too, however accomplished a worker he may become, falls straightway into the same American passion of unrest, and becomes presently the dispenser of lager bier, or a forager "mit Sigel."

There is then no American class of farm workers in the market—certainly not in the Eastern markets. The native, if he possess rural instincts, is engrossed, as I have said, with some homestead of his own, or is trying his seed-cast among the Mormons, or on the prairies. All other parties bring only a divided allegiance, and a kind of makeshift adhesion to the business; in addition to which, they bring an innocency that demands the supervision of a good farm teacher.

Such a teacher your foreman may be, or he may

not be; if the latter, and he have no capacity to convert into available workers, such motley materials, the sooner you discharge him the better; but if he have this capacity, and is, besides, so far cognizant of your ownership, as not to take offence at your presence, and to permit of your suggestions—cherish him; he has rare virtues.

From the hints I have already dropped in regard to the qualities and characteristics of the available "milkmaids" and ploughmen, it will naturally be inferred that I would not be anxious to entertain a large squad of such, under the low-browed ceilings of the country home I have described.

And here comes under observation that romanticism about equality of condition and of tastes, which many kindly and poetically-disposed persons are inclined to engraft upon their ideal of the farm life. There is, indeed, a current misjudgment on this head, which is quite common, and which the exaggerated tone of rural literature generally, from Virgil down, has greatly encouraged. The rural writers dodge all the dirty work of the farm, and regale us with the odors of the new-mown hay. The plain truth is, however, that if a man perspires largely in a cornfield of a dusty day, and washes hastily in the horse trough, and eats in shirt sleeves that date their cleanliness three days back, and loves fat pork and cabbage "neat," he will not prove the Arcadian companion at dinner, which readers of Somerville imagine,—neither on the score of conversation, or of transpiration. Active, every-day farm labor is certainly not congruous with a great many of those cleanly prejudices which grow out of the refinements of civilization. We must face the bald truth in this matter; a man who has only an hour to his nooning, will not squander it upon toilet labors; and a long day of close field-work leaves one in very unfit mood for appreciative study of either poetry or the natural sciences.

The pastoral idea,—set off with fancies of earthen bowls, tables under trees, and appetites that are sated with bread and milk, or crushed berries and sugar, and with the kindred fancies of rural swains, who can do a good day's work and keep their linen clean,—is all a most wretched phantasm. Pork, and cabbage, and dirty wristbands, are the facts.

Plainly, the milkmaids must have a home to themselves, where no dreary etiquette shall oppress them. This home, which is properly the farmer's, lies some eighth of a mile southward, upon the same highway that passes my door. For a few rods the road keeps upon a gravelly ridge, from which, eastward, stretches away the wide-reaching view I have already noted; and westward, in as full sight, is the little valley lawn, where the shadows of the copses lie splintered on the green. So it is, for a breathing space of level; then the gravelly road makes sudden plunge under a thicket of trees; a rustic culvert is crossed, which is the wasteway of the pool at the foot of the lawn; and opposite on a gentle lift of turf, all overshadowed with trees, is the farmery. Here, as before described, were outlying sheds, and leaning gables, and a wreck of castaway ploughs and carts; and the scene alive with the cluck of matronly hens, conducting broods of gleesome chickens, and with the sidelong waddle of a bevy of ducks. I have a recollection, too, of certain long-necked turkeys, who eyed me curiously on my first visit, with an oblique twist of their heads, and of a red-tasselled Tom, who sounded a gobble of alarm, as I marched upon the premises, and met me with a formidable strut. These birds are very human. I never go to the town but I see men who remind me of the gobblers; and I never see my gobblers but they remind me pleasantly of men in the town.

Immediately beyond the gates, which opened upon the farmery, was a quaint square box of red trimmed off with white (whose old-fashioned coloring I maintain), being a tenant house of most venerable age, and standing in the middle of a wild and ragged garden. The road has made two easy curves up to this point, and skirts a great hill that rises boldly on the right; on the left, and beyond the red tenant house with its clustering lilacs, and shading maples, is a mossy orchard; and with the mossy orchard on the left, and the sudden hills piling to the right, the border of the land is reached.

The wooden farmhouse, which lay so quietly under the trees, at the foot of the hill, when I first saw the place, is long since burned and gone. It was the old story of ashes in a wooden kit—very lively ashes, that one night kindled the kit, and thence spread to the shed, and in a moment half the house was in flame. It was a picturesque sight from my window on the hill; but not a pleasant one. A wild, sweeping, gallant blaze, that wrapped old powder-post timbers in its roar, and licked through crashing sashes, and came crinkling through the roof in a hundred wilful jets, and then lashed and overlaid the whole with a tent of vermilion, above which there streamed into the night great, yellow, swaying pennants of flame. But the burnt house is long since replaced by another. It would have been a simple and easy task to restore it as before: a few loads of lumber, the scheme of some country joiner, and the thing were done. But I was anxious to determine by actual trial how far the materials which nature had provided on the farm itself, could be made available.

The needed timber could, of course, be readily

obtained from the farm wood; and from the same source might also be derived the saw logs for exterior covering. But from the fact that pine is very much more suitable and durable for cover, than the ordinary timber of New England woods, the economy of such a procedure would be very doubtful; nor would it demonstrate so palpably and unmistakably, as I was desirous of doing, that the building was of home growth. I had seen very charming little farmhouses on the Downs of Hampshire, made almost entirely from the flints of the neighboring chalkbeds; and in Cumberland and Westmoreland very substantial and serviceable cottages are built out of the rudest stones, the farm laborers assisting in the work. Now, there were, scattered along the roadside, as along most country roadsides of New England, a great quantity of small, ill-shapen stones, drawn thither in past years from the fields, and serving only as the breeding ground for pestilent briars. These stones I determined to convert into a cottage.

Of course, if such an experiment should involve a cost largely exceeding that of a simple wooden house of ordinary construction, its value would be partially negatived; since I was particularly anxious to demonstrate not only the possibility of employing the humblest materials at hand, but also of securing durability and picturesqueness in conjunction with a rigid economy.

I need not say to any one who has attempted a similar task, that the builders discouraged me: the stones were too round or too small; they had no face; but I insisted upon my plan—only yielding the use of bricks for the corners, and for the window jambs.

I further insisted that no stone should be touched with a hammer; and that, so far as feasible, the mossy or weather sides of the stones should be exposed. The cementing material was simple mortar, made of shell lime and sharp sand; the only exception being one course of five or six inches in depth, laid in water cement, six inches above the ground, and intended to prevent the ascent of moisture through the mason work. The house walls were of the uniform height of ten feet, covered with a roof of sharp pitch. The gables were carried up with plank laid on vertically, and thoroughly battened; and to give picturesque effect as well as added space upon the garret floor, the gables overhang the walls by the space of a foot, and are supported by the projecting floor beams, which are rounded at their ends, but otherwise left rough. This feature, as well as the sharp pent roof, was an English one, and a pleasant reminder of old houses I had seen in the neighborhood of Gloucester.

To avoid the expense of a great number of window jambs, which, being of brick, were not of home origin, I conceived the idea of throwing two or three windows into one; thus giving, for purely economic reasons, a certain Swiss aspect to the building, and a pleasant souvenir of a sunny Sunday in Meyringen. These broad windows, it must be observed, have no cumbrous lintels of stone—for none such were to be found upon the farm; but the superincumbent wall is supported by stanch timbers of oak, and these disguised or concealed by little protecting rooflets of plank. Thus far, simple economy governed every part of the design; but to give increased architectural effect, as well as comfort, a porch, with peak corresponding in shape to the gable, was thrown out over the principal door to the south; and this porch was constructed entirely, saving its roof, of cedar unstripped of its bark. If it has not been removed, there is a parsonage house at Ambleside in the lake country of Westmoreland, which shows very much such another, even to the diamond loophole in its peak.

Again, the chimneys, of which there are two, instead of being completed in staring red, were carried up in alternate checkers of cobbles and brick, the whole surmounted by a projecting coping of mossy stones. In view of the fact that this architectural device demanded dexterous handling, I cannot allege its economy; but its extra cost was so trifling, and its pleasant juxtaposition of tints was so suggestive of the particolored devices that I had seen on the country houses of Lombardy, that the chimneys have become cheap little monuments of loiterings in Italy.

The plank of the gables, wholly unplaned, has been painted a neutral tint to harmonize with the stone, and the battens are white, to accord with the lines of mortar in the wall below; the commingled brick and stone of the house, are repeated in the chimneys above; the roof has now taken on a gray tint; the lichens are fast forming on the lower stones; a few vines,—the Virginia creeper chiefest (Ampelopsis Hederacea),—are fastening into the crevices, making wreaths about the windows all the summer through, and in autumn hang flaming on the wall. There is a May crimson, too, from the rose-bushes that are trailed upon the porch. It is all heavily shaded; a long, low wall of gray, lighted with red-bordered embrasures, taking mellowness from every added year; there are no blinds to repair; there is but little paint to renew; it is warm in winter; it is cool in summer; vines cling to it kindly; the lichens love it; I would not replace its homeliness with the jauntiest green-blinded house in the country.

Of course so anomalous a structure called out the witticisms of my country neighbors. "Was it a blacksmith's shop?" "Was it a saw mill?" and with a loud appreciatory "guffaw" the critics pass by.

Our country tastes are as yet very ambitious; homeliness and simplicity are not appetizing enough. But in time we shall ripen into a wholesome severity, in this matter. I am gratified to perceive that the harshest observers of my poor cottage in the beginning, have now come to regard it with a kindly interest. It mates so fairly with the landscape,—it mates so fairly with its purpose; it is so resolutely unpretending, and carries such air of permanence and durability, that it wins and has won upon the most arrant doubters.

The country neighbors were inclined to look upon the affair as a piece of stupidity, not comparable with a fine white house, set off by cupola and green blinds. But it was presently observed that cultivated people from the town, in driving past, halted for a better view; the halts became frequent; it was intimated that So-and-so, of high repute, absolutely admired the homeliness. Whereupon the country critics undertook an inquiry into the causes of their distaste, and queried if their judgment might not have need of revision. Did their opinion spring from a discerning measurement of the real fitness of a country house, or out of a cherished and traditional regard for white and green?

The final question, however, in regard to it, as a matter of practical interest, is one of economy. Can a house of the homely material and character described be built cheaply? Unquestionably. In my own case the cost of a cottage fifty feet by twenty-six, and with ten-feet walls—containing five serviceable rooms, besides closets on its main floor, and two large chambers of good height under the roof, as well as dairy room in the east end of the cellar—was between eleven and twelve hundred dollars. The estimates given me for a wooden house, of the stereotyped aspect and similar dimensions, were within a few dollars of the same sum.

It must be remembered, however, that any novelty of construction in a particular district, costs by reason of its novelty; the mason, too, charges for the possible difficulties of overcoming his inexperience in the material. The carpenter rates the rough joining at the same figure with the old mouldings and finishing boards, to which he is accustomed, and of which he may have a stock on hand. Yet, notwithstanding these drawbacks, the work was accomplished within the limits of cost which the most economic would have reckoned essential to a building of equal capacity.

It is further to be considered that while I paid skilful masons for this rough work the same price which they exacted for the nice work of cities, it would have been quite possible for an intelligent proprietor to commit very much of it to an ordinary farm laborer, and so reduce the cost by at least one third.

I have dwelt at length upon this little architectural experience, because I believe that such meagre details of construction as I have given may be of service to those having occasion to erect similar tenant houses; and again, because in view of the fact that we must in time have a race of farm laborers among us, who shall also be householders, I count it a duty to make such use of the homely materials at hand, as shall insure durability and comfort, while the simplicity of detail will allow the owner to avail himself of his own labor and ingenuity, in the construction.

A Sunny Frontage.

SUCH a farmhouse as I have described, should have, in all northern latitudes, a sheltered position and a sunny exposure. Of course, a situation convenient to the fields under tillage, and to other farm buildings, is to be sought; but beyond this, no law of propriety, of good taste, or of comfort, is more imperative than shelter from bleak winds, and a frontage to the south. No neighbor can bring such cheer to a man's doorstep as the sun.

There are absurd ideas afloat in regard to the front, and back side of a house, which infect village morals and manners in a most base and unmeaning way. In half the country towns, and by half the farmers, it is considered necessary to retain a pretending front side upon some dusty street or highway, with tightly closed blinds and bolted door; with parlors only ventured upon in an uneasy way from month to month, to consult some gilt-bound dictionary, or Museum, that lies there in state, like a king's coffin. The occupant, meantime, will be living in some back corner,—slipping in and out at back doors, never at ease save in his most uninviting room, and as much a stranger to the blinded parlor, which very likely engrosses the best half of his house, as his visitor, the country parson. All this is as arrant a sham, and affectation, as the worst ones of the cities.

It is true that every man will wish to set aside a certain portion of his house for the offices of hospitality. But the easy and familiar hospitalities of a country village, or of the farmer, do not call for any exceptional stateliness; the farmer invites his best friends to his habitual living room; let him see to it then that his living room be the sunniest, and most cheerful of his house. So, his friends will come to love it, and he, and his children—to love it and to cherish it, so that it shall be the rallying point of the household affections through all time. No sea so distant, but the memory of a cheery, sunlit home-room, with its pictures on the wall, and its flame upon the hearth, shall haunt the voyager's thought; and the flame upon the hearth, and the sunlit window, will pave a white path over the intervening waters, where tenderest fancies, like angels, shall come and go. No soldier, wounded on these battle fields of ours, and feeling the mists of death gathering round him, but will call back with a gushing fondness such glimpse of a cheery and cherished hearthstone, and feel hope and heart lighted by the vision—bringing to his last hold on earth his most hallowed memories; and so, binding by the tenderest of links, the heartiest of the Old life, to the bloody dawn of the New.

There is a deeper philosophy in this than may at first sight appear. Who shall tell us how many a breakdown of a wayward son, is traceable to the cheerless aspect of his own home, and fireside?

But just now I am no moralist—only housebuilder. In the farm cottage, whose principal features I have detailed, I have given fifty feet of frontage to the south, and only the gable end, with its windows, to the street. As I enter the white wicket by the corner, under the elm tree which bowers it, the distribution counts thus: a miniature parlor with its lookout to the street, and a broad window to the south;

next is the rustic porch, and a door opening upon the hall; next, a broad living-room or kitchen, with its generous chimney, and this flanked by a wash room, or scullery, from which a second outer door opens upon the southern front. To this latter door, which may have its show of tubs, tins, and drying mops, a screen of shrubbery gives all needed privacy from the street, and separates by a wall of flowering things from the modest pretensions of the entrance by the porch. At least, such was an available part of the design. If the good woman's poultry, loving so sunny a spot, will worry away the rootlets of the lower flowering shrubs, and leave only a tree or two for screen, it is an arrangement of the leafy furniture, over which the successive occupants have entire control. The noticeable fact is, that the best face of the cottage, and its most serviceable openings, whether of window or door, are given to the full flow of the sun, and not to the road side. What is the road indeed, but a convenience? Why build at it, or toward it, as if it were sovereign, or as if we owed it a duty or a reverence? We owe it none; indeed, under the ordering of most highway surveyors, we owe it only contempt. But the path of the sun, and of the seasons, is of God's ordering; and a south window will print on every winter's morning a golden prayer upon the floor; and every summer's morning the birds and bees will repeat it, among the flowers at the southern door.

Farm Buildings.

HAVING looked after the farm cottage, I come now to speak of the equally homely subject of barns and outbuildings. Of these, such as they were, I found abundance upon the premises, standing at all imaginable angles, and showing that extraordinary confusion of arrangement for which many of our old-fashioned farmers have a wonderful aptitude. Should they all be swept away, and a new company of buildings erected? The stanch timbers and the serviceable condition of many of them forbade this, as well as considerations of prudence. Besides which, I have no admiration for that incongruity which often appears at the hands of those who are suddenly smitten with a love for the country—of expensive and jaunty farm architecture in contrast with a dilapidated farm. I believe in a well-conditioned harmony between farm products and the roofs that shelter them, and that both should gain extent and fulness, by orderly progression. It has chanced to me to see here and there through the country very admirable appliances of machinery and buildings, which, on the score of both cost and needfulness, were out of all proportion to the fertility and the order of the fields. I see, too, not unfrequently, very showy palings in the neighborhood of a country house, which are flanked by the craziest of slatternly fences; whereat it always occurs to me, that the expenditure would be far better distributed in giving a general neatness and effectiveness to all the enclosures, rather than lavished upon a little spurt of white splendor about the house. A fertility too gross for the buildings, so as to bubble over in ricks and temporary appliances, is to me a far more cheery sight agriculturally, than buildings so grand as utterly to outmatch and overshadow all productive capacity of the land. A kernel too big for the nut, promises to my taste a better relish than a nut too big for the kernel.

These seem to me, at the worst, very plausible reasons, if there had been no final, prudential ones, for making the best of the old buildings at hand—by re-arrangement, new grouping, and by shutting up such gaps between the disjointed parts, as should reduce the whole to a quadrangular order, and offer sunny courts for the cattle.

If a sunny exposure, and grateful shelter from harsh winds be good for the temper of the farm wife and her household, they are even better for all the domestic animals; and it is an imperative condition of the arrangement of all farm buildings in our climate that they offer a sheltering lee, and have their principal openings, specially of windows, to the south. Protection against summer heats, if needed for stalled animals, it is easy to supply; but an equivalent for the warmth of the winter's sun, I know no name for.

Another condition of all judicious arrangement, which is even more important, is such disposition of the yards and cellars, as shall prevent all waste of manurial resources of whatever kind, whether by undue exposure, or by leakage. And in this connection, I may mention that it is a question seriously mooted, and worthy of full investigation—if the fertilizing material of a farm will not warrant special shelter as fully as the crops. All experience certainly confirms the fact that such as is taken from under cover, provided only the moisture is sufficient, is worth the double of that which has been exposed to storms. What chemical laws relating to agriculture confirm this fact, I may have occasion to speak of in another chapter; at present I note only the results of practical observation, without reference to underlying causes.

The books would have recommended me to construct an extensive tank, to which drains should conduct all the wash from the courts and stables. But this would involve water carts, and other appliances, liable to injury under rough handling; besides demanding a nicety of tillage, and a regularity of distribution which, at first, could not be depended on. That the liquid form is the one, under which manurial material under a complete system of culture, will work the most magical results, I have no doubt. But until that system is reached, very much can be done in the way of economizing the fertilizing elements of the farmyard, short of the tank and the water cart; and this by modes so simple, and at an expense so small, as to be within the reach of every farmer.

Let me illustrate, in the plainest possible manner, by my own experience. The barn, as I have said, was slatternly; it had yielded a little to the pinching northwesters, and by a list (as seamen say) to the southeast, gave threat of tumbling upon the cattle yard. This yard lay easterly and southerly, in a ragged, stony slope, ending on its eastern edge with a quagmire, which was fed by the joint wash of the yard and the leakage of a water trough supplied from a spring upon the hills. The flow from this quagmire, unctuous and fattening, slid away down a long slope into the meadow,—at first so strong, as to forbid all growth; then feeding an army of gigantic docks and burdocks; and after this giving luxuriant growth to a perch or two of stout English grass. But it was a waste of wealth; it was like a private, staggering under the rations of a major-general. I cut off the rations. With the stones which were in and about the yard, I converted its lazy slope into two level courts; and so arranged the surface, that the flow from the upper should traverse the lower one; from which, in turn, the joint flow of fertilizing material fell though a few tiles in the lower terrace wall, upon the head of a long heap of compost, which was ordered to be always replaced, as soon as removed. The leakage of the water trough being cured, its overflow was conducted to the pasture below, where the second overflow,—for the stream was constantly running,—would do no injury, and would be available as a foraging mudpool for the ducks. By this simple re-adjustment of surface, and of the water flow, I have no question but I fully doubled the yearly value of the manures.

I still further mended matters by carrying the cow stables along all the northern frontier of the yard, in such sort as to afford an ample sunny lee; and extending this new pile of building over the eastern terrace wall, I gained an open cellar below for my store pigs, who range over the ground where the burdocks so thrived before, with occasional furtive examinations of the compost heap which receives the flow from above.

I do not name this disposition of buildings and of surfaces, as one to be copied, or as the one which I should have chosen to make, in the event of a thorough reconstruction; but only as one of those simple, feasible improvements of the old conditions which are met with everywhere; improvements, moreover, which involve little or no cost, beyond the farmer's own labor, and no commitment to the theories of Mechi or of Liebig. A ragged-coated man should be grateful for a tight bit of linsey-wolsey to his back, until such time as he comes to the dignity of broadcloth.

Four fifths of those who undertake farming,—not as an amusement or simply as an occupation, but as the business of their life, and upon whom we are dependent for our potatoes, veal, and cider (to say nothing more),—are compelled to do the best they can with existing buildings; and Stephens' plans of a 'farmsteading' are as much Greek to them, as the 'Works and Days' of Hesiod. A hint, therefore, of judicious adaptation of old buildings, may be all they can digest with that practical relish with which a man accepts suggestions that are within the compass of his means and necessities.

Again, the British or Continental needs in the matter of farm constructions, are totally different from American need, in all northern latitudes. The British farmer can graze his turnips into January; and I have seen a pretty herd of Devon cows cropping a fair bite of grass, under the lee of the Devon Tors, into February. We, on the contrary, have need to store forage for at least six months in the year. Hay begins to go out of the bays with the first of November at the latest, and there is rarely a good bite upon the pastures until the tenth of May. For this reason there is required a great breadth of barn room.

The high cost of labor, too, forbids that distribution of the farm offices over a considerable area of surface, which is characteristic of the British steading. The tall buildings, which are just now so much in vogue with enterprising American farmers, situated by preference upon swiftly sloping land, and giving an upper floor for forage, a second and lower one for granary and cattle, and a third for manure pit, have been suggested and commended chiefly for their great economy of labor; one man easily caring for a herd, under these conditions of lodgment, which upon the old system would demand two or three.

Machinery, too, which must presently come to do most of the indoor work upon a well-managed farm of any considerable size, will require for its effective service compact buildings.

Let me repeat the conditions of good American barns. They must suffice for ample protection of the harvested crops; ample and warm shelter for the animals; security against waste of manurial resources; and such compactness of arrangement as shall warrant the fullest economy of labor. With these ends reached, they may be old or new, irregular or quadrangular—they are all that a good farmer needs in the way of architecture, to command success.

The Cattle.

"WHAT sort o' cattle d'ye mean to keep, Squire?" said one of my old-fashioned neighbors, shortly after my establishment. "Squire" used to be the New England title for whatever man, not a clergyman or doctor, indulged in the luxury of a black coat occasionally, upon work days. But in these levelling times, I am sorry to perceive that it is going by; and I only wear the honor now, at a long distance from home, in the 'up-country.'

To return to the cattle; my neighbor's question was a pertinent one. Not what cattle did I admire most, or what cattle I thought the finest; but what cattle shall I keep?

In this, as in the matter of the house, of the outbuildings and of the roadway, I believe thoroughly in adaptation to ends in view. If I had been undertaking the business of a cattle breeder, I should have sought for those of the purest blood, of whatever name; if I had counted upon sales to the butcher, my choice would have been different; if, again, butter had been the aim, I am sure I should have made no great mistake in deciding for the sleek Jersey cattle. But for mere supply of milk, under ordinary conditions of feeding, I do not know that any breed has as yet established an unchallenged claim to the front rank. The Devons, Ayrshires, and Shorthorns, each have their advocates; for the latitude and pasturage of New England, if I were compelled to choose between the three, I should certainly choose the Ayrshires; but I am satisfied that a more successful milk dairy can be secured by a motley herd of natives, half-bloods, and animals of good promise for the pail, than by limitation of stock to any one breed. I am confirmed in this view by the examples of most large dairies of this country, as well as by many in Great Britain. I particularly remember a nice little herd which I had the pleasure of seeing, some years since, at the excellently managed farm of Glas-Nevin in the environs of Dublin: sleek animals all, and thoroughly cared for; but showing a medley of races; the queen milker of all—as it chanced—having lineage in which the Ayrshire, the Shorthorn and Devon were all blended.

I know there are very many cattle fanciers, and stanch committee men, who will not approve this method of talking about mixed stables, and of a medley of different races,—as if a farmer were at liberty to make his choice of cattle, with the same coolness with which he would make his choice of ploughs or wagons, and to tie up together, if the humor takes him, animals which the breeders have been keeping religiously apart for a few score of years.

But I do not share in this punctiliousness. I believe that these animals all, whether of the Herd-book or out of it, must be measured at last, not by their pedigree or title, but by their fitness for humble farm services. A family name may be a good enough test of any animal—biped or other—from whom we look for no particular duty, save occasional exhibition of his parts before public assemblages; but when our exigencies demand special and important service, we are apt to measure fitness by something more intrinsic.

The cattle breeders are unquestionably doing great benefit to the agricultural interests of the country; but the essential distinction between the aims of the breeder and farmer should not be lost sight of. The first seeks to develop, under the best possible conditions of food and shelter, those points in the animal which most of all make the distinction of the race. The farmer seeks an animal, or should, which in view of climate, soil, and his practice of husbandry, shall return him the largest profit, whether in the dairy, under the yoke, or in the shambles. He has nothing to do with points, but the points that shall meet these ends. There is no reason why he should limit himself to one strain of blood, unless that strain meets and fills every office of his farm economy, any more than he should narrow his poultry range to pea-fowl, or to golden pheasant.

I think I may have talked somewhat in this strain to my old neighbor, who asked after the "squire's cattle"—but not at such a length; and I think that he offered some such corollary as this:

"Squire, them English cows is handsome critturs enough to look at; but ye have to keep a follerin' on 'em up with a meal tub."

It is very easy to lay down a charming set of rules for the establishment of a good herd (and for that matter—of a good life); but—to follow them?

I will be bound to say that there was never a prettier flock of milch cows gathered in any man's stables than the superior one which I conjured up in my fancy, after an imaginative foray about the neighborhood. But it was not easy to make the fancy good. Mr. Flint, in his very capital book upon milch cows and dairy farming, gives a full elucidation of that theory of M. Guénon, by which the milking properties of an animal can be determined by what is called the escutcheon,—being certain natural markings, around the udder upon the inner parts of the thighs. It is perhaps needless to say, that such minute observation as would alone justify a decision based upon this theory, might sometimes prove awkward, and embarrassing. Upon the whole, I should counsel young farmers in summer clothing, and away from home, to judge of a cow by other indicia.

Still, the theory of M. Guénon* has its value; and I am persuaded that he was worthily adjudged the gold medal at the hands of the Agricultural Society of Bordeaux. But with this, and all other aids—among which I may name the loose preëmptory reflections and suggestions of certain adjoining farmers—I was by no means proud of the appearance of the little herd of twelve or fourteen cows with which operations were to commence.

The popular belief, that all jockeyism and cheatery is confined to horse dealings, is too limited. Whoever will visit the cow stables in Robinson street, or near to Third avenue, upon a market day, may observe a score or two of animals with painfully distended udders (the poor brutes have not been milked in the last forty-eight hours), throwing appealing glances about the enclosure, and eyeing askance certain bullet-headed calves, which are tied in adjoining stalls, but which

* The interested agricultural reader may consult "*Choix des Vaches Laitières, par M. Magne, Paris*," for full exhibition of the system.

have no more claim upon the maternal instincts of the elder animals, than the drovers themselves. It is all a bald fiction; the true offspring have gone to the butchers months ago; and if the poor, surcharged brutes accept of the offices of the little staggering foundlings, it is with a weary poke of the head, that is damning to the brutality of the drovers.

It would be too much to say that I have never been deceived by these people; too much to say that honest old gentlemen of innocent proclivities did never pass upon me certain venerable animals, with the tell-tale wrinkles rasped out of their horns. One of this class, of a really creditable figure, high hip bones, heavy quarters, well marked milk veins, I was incautious enough to test by a glance into her mouth. Not a tooth in her old head!

I looked accusingly at the rural owner, who was quietly cutting a notch in the top rail of his fence.

"Waal, yes—kinder rubbed off; but she bites pooty well with her gooms."

Among the early purchases, and among the animals that promised well, was a dun cow, which it was found necessary, after a few weeks of full feeding, to cumber with a complicated piece of neck furniture, to forbid her filching surreptitiously what properly belonged to the pail. Self-milkers are not profitable. I have faith in the doctrine of rotation, and the quick reconversion of farm products into the elements of new growth. But here was a case of reconversion so rapid, as to be fatal to all the laws of economy. It suggested nothing so strongly, as that rapid issue of government money, which finds immediate absorption among the governmental officials. Does the government really milk itself; and can no preventive be found in the way of neck machinery, or other?

Another animal was admirable in every point of view; I found her upon one of the North River wharves, and the perfect outline of her form and high-bred action, induced a purchase, even at a long figure; but the beast proved an inveterate kicker.

The books recommend gentleness for the cure of this propensity; so does humanity; I concurred with both in suggesting that treatment to Patrick.

"Gintle is it? And bedad, sir, she's too ould for a cure. I'm thinking we must tie her legs, sir; but if ye orders it, bedad, it's meself can be gintle.

"Soh, Moolly—soh—soh (and a kick); soh, ye baste (a little livelier), soh (and a kick)—soh, blast ye!—*soh*, Moolly—SOH, Katy—SOH (and a crash); och, you ould baste ye—take that!" and there is a thud of the milking stool in the ribs.

The "gintleness" of Patrick is unavailing. But the cow is an excellent animal, and not to be hastily discarded. Milker after milker undertook the conquest, but with no better success. The task became the measure of a man's long-suffering disposition; some gave over, and lost their tempers before the first trial was finished; others conjured down the spirit by all sorts of endearing epithets and tenderness, until the conquest seemed almost made; when suddenly pail, stool, and man would lapse together, and a stream of curses carry away all record of the tenderness. We came back at last to Patrick's original suggestion; the legs must be tied. A short bit of thick rope passed around one foot and loosely knotted, then passed around the second and tied tightly in double knot, rendered her powerless. There was a slight struggle, but it was soon at an end; and she made no opposition to the removal of the thong after the milking was over. With this simple provision, the trouble was all done away; and for a whole year matters went well. But after this, there came a reformer into control of the dairy. The rope was barbarous; he didn't believe in such things; he had seen kicking cows before. A little firmness and gentleness would accomplish the object better; God didn't make cows' legs to be tied. The position was a humane one, if not logical. And the thong was discarded.

"Well. Patrick," said I, two days after, "how fares the cow?"

"And begorra, it's the same ould baste, sir."

A few days later I inquired again after the new regimen of gentleness and firmness.

"Begorra," said Patrick, "she's kicked him again!"

A week passed; and I repeated the inquiries.

"Begorra, she's kicked him *again!*" screamed Patrick; "and it's a divil's own bating he's been giving the ould baste."

Sure enough, the poor cow was injured sadly; her milking days were over; and in a month she went to the butcher. And this advocate of gentleness and firmness was one of the warmest and most impassioned philanthropists I ever met with.

The moral of the story is,—if a cow is an inveterate kicker, tie her legs with a gentle hand, or kill her. Beating will never cure, whether it come in successive thuds, or in an explosive outbreak of outrageous violence. I suspect that the same ruling is applicable to a great many disorderly members of society.

Although the cases I have cited were exceptional, and although my little herd had its quiet, docile, profit-giving representatives, yet I cannot say that it was altogether even with my hopes or intentions.

Two stout yoke of those sleek red cattle, for which southern New England is famous, had their part to bear in the farm programme, besides a sleek young Alderney bull, and a pair of sturdy horses. There were pigs with just enough of the Suffolk blood in them to give a shapely outline, and not so much as to develop that red scurfy baldness, which is to my eye rather an objectionable feature of high breeding and feeding—whether in men or pigs.

Ducks, turkeys, and hens, in a fluttering brood, brought up the rear. With these all safely bestowed about the farm buildings which I have briefly indicated, and with a rosy-nosed, dapper little Somersetshire man, who wore his tall Sunday-beaver with a slight cant to one side, established as lieutenant-governor in the cottage, the reins seemed fairly in hand.

CROPS AND PROFITS.

III.

CROPS AND PROFITS.

The Hill Land.

BEFORE we keep company farther—the reader and I—let me spread before him, as well as I may, a map of the farm land. I may describe it, in gross, as a great parallelogram, intersected by the quiet public highway, which divides it into two great squares. The eastern square is, for the most part, as level as the carpet on my library floor, and its crops make checkers like the figures on the ingrain. The eastern half is toward the town; and upon its edge, by the highway, are the farm buildings I have grouped around the stone cottage. The western half is rolling; and beyond the whitey-gray farm-house, with which I entered upon my portraiture, it heaves up into a great billow of hill, half banded with woodland, and half green with pasture.

This billow of hill, dipped down between my home and the stone cottage, into a little valley, which I have transmuted, as before described, into a lawn of grass land, with its clumps of native trees and flowering shrubs, and its little pool, under the willows, that receives the drainage. Elsewhere, beyond, and higher, its surface was scarred with stones of all shapes and sizes; orderly geology would have been at fault amid its debris;—there were boulders of trap, with clean sharp fissures breaking through them;—there were great flat fragments of gneiss covered with gray lichens;—there were pure granitic rocks worn round,—perhaps by the play of some waves that have been hushed these thousand years; and there were exceptional fragments of coarse red sandstone, frittered half away by centuries of rain, and leaving protruding pimples of harder pebbles. In short, Professor Johnston, who advised (in Scotland) the determination of a farm purchase by the character of the subjacent and adjoining rocks, would have been at fault upon my hillside. A short way back, amid the woods, he would have found a huge ridge of intractable serpentine; the boulders he would have discovered to be of most various quality; and if he had dipped his spade, aided by a pick, he would have found a yellow, ferruginous conglomerate, which the rains convert into a mud that is all aflow, and which the suns bake into a surface, that with the sharpest of mattocks would start a flood of perspiration, before he had combed a square yard of it into a state of garden pulverization.

Lying above this, however, was a vegetable mould, with a shiny silicious intermixture (what precise people would call a sandy loam), well knitted together by a compact mass of the roots of myrtles, of huckleberry bushes, and of ferns. Geologically, the hill was a 'drift;' agriculturally, considering the steep slopes and the matted roots, it was uninviting; pictorially, it was rounded into the most graceful of cumulated swells, and all glowing with its wild verdure; practically, it was a coarse bit of neglected cow-pasture, with the fences down, and the bushes rampant.

What could be done with this? It is a query that a great many landholders throughout New England will have occasion some day to submit to themselves, if they have not done so already. Overfeeding with starveling cows, and a lazy dash at the brush in the idle days of August, will not transform such hills into fields of agricultural wealth. Under such regimen they grow thinner and thinner. The annual excoriation of the brush above ground, seems only to provoke a finer and firmer distribution of the roots below; and the depasturing by cows—particularly of milch animals, folded or stalled at night—will gradually and surely diminish the fertilizing capital of such grazing land. It is specially noticeable that the deterioration under these conditions, is much more marked upon hill lands than upon level meadows.*

In the back country, such old pastures with their brush and scattered stones, will feed sheep profitably, and will grow better under the cropping. But in the immediate neighborhood of towns, where every barkeeper has his half dozen dogs, and every Irish family their cur, and every vagabond his canine associate, sheep can only be kept at a serious risk of immolation for the benefit of these worthies. Proper legislation might interpose a bar, indeed, to such sacrifice of agricultural interests,—if legislation were not so largely in the hands of dog-fanciers.

The sheep are not the only sufferers.

Shall the hill be ploughed? It is not an easy task to lay a good furrow along a slope of forty-five degrees, with its seams of old wintry torrents, its occasional boulders, and its matted myrtle roots; and, if fairly accomplished, the winter's rains may drive new seams from top to bottom, carrying the light mould far down under walls, and into useless

* This is perhaps more apparent than real, from the fact that upon level lands the droppings are more evenly distributed.

places,—leaving harsh yellow scars, that will defy the mellowest June sunshine.

A city friend, with city aptitude, suggests—terraces; and instances the pretty ones overhung with vines, which the traveller may see along the banks of the Rhine.

I answer kindly; and in the same vein—suggest that such scattered rocks, as are not needed, may be thrown into the shape of an old watch tower—with Bishop Hatto's for a model—to mimic the Rhine ruins.

—— "Charming! and when the grapes are ripe, drop me a line." And my city friend plucks a bit of penny-royal, and nips it complacently.

Terracing might be done in a rude but substantial way, at the cost of about fifteen hundred dollars the acre. This might do at Johannisberg; but hardly, in a large way, in Connecticut. Crops must needs be exceeding large upon such terraces, to compete successfully with those of a thriving 'forehanded' man, who farms upon a land capital of less than a hundred dollars to the acre.

I abandoned the design of terraces. And yet, there are times when I regale myself for hours together, with the pleasant fancy of my city friend. His terraces should be well lichened over now; and I seem to see brimming on the successive shelves of the hill, great festoons of vines, spotted with purple clusters; amidst the foliage, there gleams, here and there, the broad hat of some vineyard dresser (as in German pictures), and crimson kirtles come and go, and songs flash into the summer stillness, and a soft purple haze wraps the scene, and thickens in the hollows of the land, and swims fathoms deep around the ruin——

"Square, what d'ye ask apiece for them suckers?"

It is my neighbor, who has clambered up, holding by the myrtle bushes, to buy a pig.

The vexed question of the proper dressing and tillage of the hillside, is still in reserve. I resolved it in this wise:—Of the rocks most convenient, and least available for fencing purposes, I constructed an easy roadway, leading by gradual inclination from top to bottom; other stones were laid up in a substantial wall, which supplies the place of a staggering and weakly fence, which every strong northwester prostrated; still others, of a size too small for any such purpose, were buried in drains, which diverted the standing moisture from one or two sedgy basins on the hill, and discharged the flow upon the crown of a gravelly slope. There, I have now the pleasure of seeing a most luxuriant growth of white clover and red top, fertilized wholly by the flow of water which was only harmful in its old locality. I next ordered, in the leisurely time of later autumn, the grubbing up of the patches of myrtles and briers, root and branch; these with the mossy turf that cumbered them, after thorough drying, were set on fire, and burned slumberously, with a little careful watching and tending, for weeks together. I was thus in possession of a comparatively smooth surface, not so far disintegrated as to be subject to damaging washes of storm, besides having a large stock of fertilizing material in the shape of ashes.

In the following spring, these were carefully spread; a generous supply of hay-seed sown, and still further, an ample dressing of phosphatic guano. The hillside was then thoroughly combed with a fine-toothed Scotch harrow, and the result has been a compact lively sod, and a richer bite for the cattle.

Again, upon one or two salient points of the hill, where there were stubborn rocks which forbade removal, I have set little coppices of native evergreens, which, without detracting in any appreciable degree from the grazing surface, will, as they grow, have charming effect, and offer such modicum of shade as all exposed pasture lands need. One who looked only to simple farm results, would certainly never have planted the little coppices, or hedged them, as I have done, against injury. But it appears to me that judicious management of land in the neighborhood of large towns, should not ignore wholly, the conservation of those picturesque effects, which at no very remote time, may come to have a marketable value, greater even than the productive capacity of the soil.

I have even had the hardihood to leave upon certain particularly intractable spots of the hill land, groups of myrtles, briers, scrubby oaks, wild grapes, and birches, to tangle themselves together as they will, in a wanton savagery of growth. Such a copse makes a round perch or two of wilderness about the sprawling wreck of an old cellar and chimney, which have traditional smack of former Indian occupancy; and the site gives color to the tradition;—for you look from it southeasterly over three square miles of wavy meadows, through which a river gleams; and over bays that make good fishing ground, and over a ten-mile reach of shimmering sea. A little never-failing spring bubbles up a few yards away; and to the westward and northward, the land piles in easy slope, making sunny shelter, where,—first on all the hillside,—the snow vanishes in Spring. The Indian people had a quick eye for such advantages of position.

In still further confirmation, I have turned up an

arrow-head or two in the neighborhood, chipped from white quartz, and as keen and sharp as on the day they were wrought.

I am aware that what are called 'tidy farmers' would have brushed away these outlying copses, no matter what roughnesses they concealed; but I suspect their rude autumn clippings with a bush-hook, would only have provoked a spread of the rootlets; and if effectual, would have given them only a bit of barrenness.

Up-country farmers are overtaken from time to time, with what I may call a spasmodic tidiness, which provokes a general onslaught with bill-hooks and castaway scythes, upon hedge rows and wayside bushes, and pasture thickets,—without considering that these thickets may conceal idle stone heaps or decrepid walls, which are as sightless as the exterminated bush; and their foray leaves a vigorous crop of harsh stubs, which, with the next season, shoot up with more luxuriance than ever, and leave no more available land within the farmer's grasp than before. Wherever it is profitable to remove such wild growth, it is profitable to exterminate it root and branch. Half doing the matter is of less worth than not doing it at all. But it is well to consider before entering upon such a campaign, if the end will justify the labor; and if the recovered strips of land will carry remunerative crops. If otherwise, let the wild growth enjoy its wantonness. It may come to be a little scattered range of wood in time, and so have its value; it may offer shelter against the sweep of winds; it will give a nursing place for the birds,—and the birds are the farmer's friends.

I am loth to believe that the natural graces of woodland and shrubbery are incompatible with agricultural interests; and a true farm economy seems to me better directed in making more thorough the tillage of the open lands, than in making Quixotic foray upon the bushy fastnesses of outlying pastures.

When a dense population shall have rendered necessary the employment of every foot of our area for food-growing purposes, it may be incumbent on us to cleave all the rocks, and to clear away all the copses: but until then, I shall love to treat with a tender consideration the green mantle—albeit of brambles and wild vines—with which Nature covers her roughnesses; and I seem to see in the streaming tendrils, and in the nodding tassels of bloom which bind and tuft these wild thickets of the hills, a sampler of vegetable luxuriance, which every summer's day provokes and defies all our rivalry of the fields.

What is called tidiness, is by no means always taste; and I am slow to believe that farm economy must be at eternal war with grace. I know well that no inveterate improver should ever tempt me to extirpate the dandelions from the green carpet of my lawn, or to cut away the wild Kalmia bush, which in yonder group among the rocks, is just now reddening into its crown of blossoms.

The Farm Flat.

IT is a different matter with the eighty acres of meadow which lie stretched out in view from my door. There, at least, it seemed to me, must be a clean, clear sweep for the furrows. Yet I remember there were long wavy lines of elder-bushes, and wild-cherries, groping beside the disorderly dividing fences. There were weakly old apple-trees, with blackened, dead tops, and with trunks half concealed by thickets of dwarfish shoots; there were triplets of lithe elms, and hickory trees, scattered here and there;—in some fields, stunted, draggled cedar bushes, and masses of yellow-weed;—a little patch of ploughed-land in the corner of one enclosure, and a waving half acre of rye in the middle of the next. The fences themselves were disjointed and twisted, —the fields without uniformity in size, and with no order in their arrangement.

"I think we must mend the look of these meadows, Coombs?"

And the dapper Somersetshire man, with his hat defiantly on one side—"Please God, and I think we will, sir."

I must do him the justice to say that he was as good as his word. In looking over the scene now, I find no straggling cedars, no scattered shoots of elms; the wayward elders, and the wild-cherries save one protecting and orderly hedgerow along the northern border of the farm—are gone. The decrepid apple-trees are rooted up, or combed and pruned into more promising shape. Ten-acre fields, trim and true, are distributed over the meadow land, and each, for the most part, has its single engrossing crop.

As I look out from my library window to-day—and the learned reader may guess the month from my description—I see one field reddened with the lusty bloom of clover, which stands trembling in its ranks, and which I greatly fear will be doubled on its knees with the first rain storm; another shows the yellowish waving green of full-grown rye, swaying and dimpling, and drifting as the idle winds will; another is half in barley and half in oats—a bristling green beard upon the first, the oats just flinging out their fleecy, feathery tufts of blossom; upon another field, are deep dark lines beneath which, in September, there are fair hopes of harvesting a thousand bushels of potatoes; yet another, shows fine lines of growing corn, and a brown area, where a closer look would reveal the delicate growth of fresh-starting carrots and mangel. All the rest in waving grass; not so clean as could be wished, for I see tawny stains of blossoming sorrel, and fields whitened like a sheet, with daisies.

If there be any cure for daisies, short of a clean fallow every second year, I do not know it; at least, not in a region where your good neighbors allow them to mature seed every year, and stock your fields with every strong wind, afresh.

Heavy topdressing is recommended for their eradication, but it is not effective; so far as I can see, the interlopers, if once established, enjoy heavy feeding. A rye crop is by many counted an exterminator of this pest; but it will find firm footing after rye. Thorough and clean tillage, with a system of rotation, afford the only security.

It is not Burns' "wee-tipped" daisy that is to be dealt with; it is a sturdier plant—our ox-eye daisy of the fields; there is no modesty in its flaunting air, and the bold uplift of its white and yellow face.

I never thought there was a beauty in it, until, on a day—years ago—after a twelvemonth's wandering over the fields of the Continent, I came upon a little pot of it, under the wing of the Madeleine, on the streets of Paris. It was a dwarfish specimen, and the nodding blossoms (only a pair of them) gave a modest dip over the edge of the red crock, as if they felt themselves in a country of strangers. But it was the true daisy for all this, and I greeted it with a welcoming franc of purchase money, and carried it to my rooms, and established it upon my balcony, where, while the flower lasted, I made a new Picciola of it. And as I watered it, and watched its green buttons of buds unfolding the white leaflets, wide visions of rough New England grasslands came pouring with the sunshine into the Paris window, and with them,—the drowsy song of locusts,—the gushing melody of Bob-o'-Lincolns,—until the drum-beat at the opposite Caserne drowned it, and broke the dream.

These living and growing souvenirs of far-away places, carry a wealth of interest and of suggestion about them, which no merely inanimate object can do. I have flowers fairly pressed, not having wholly lost their color, which I plucked from the walls of Rome, and others from a house-court of the buried Pompeii; but they are as dead as the guide books that describe the places.

It is different wholly with a little potted Ivy which a friend has sent from the walls of Kenilworth. It clambers over a rustic frame within the window—a tiny, but a real offshoot of that great mass of vegetable life which is flaunting over the British ruin; a little live bubble as it were, from that stock of vitality which is searching all the crannies of the masonry that belongs to the days of Elizabeth.

I never look at it in times of idle musing, but its shiny leaflets seem to carry me to the gray wreck of castle: and the tramp through the meadows from Leamington comes back—the wet grass, the gray walls, the broad-hatted English girls, hovering with gleeful laughter about the ruin, and the flitch of bacon hanging in the gatekeeper's house. Other-times, the dainty tendrils of the vine lead me still farther back; and Leicester, Amy Robsart, Essex, and Queen Bess with her followers, and all her court,—come trooping to my eye in the trail of this poor little exiled creeper from Kenilworth.

But this is not farming.

"Coombs," said I, "what shall we plant upon the flat?"—not that I had no opinion on the subject, but because in farming, there is a value in the suggestions of every practical worker.

The Somersetshire man leans his head a little, as if considering:—"We must have some *artificial*, sir, —for the cows—Mangel or pale Belgians,—both good, sir; some oats for the 'osses, sir; potatoes, sir, is a tidy crop—"

I observe that Englishmen and Scotchmen are disposed to slight our standard crop of maize. They do not understand it. They fail of making a creditable show in comparison with the old-school native farmers, who, by dint of long experience, have acquired the habit (rather habit than capacity) of making a moderate crop of corn with the least possible amount of tillage and of skill. To turn over a firm grass sward, and plant directly upon the inverted turf, without harrowing, or ridging, or drilling, is contrary to all the old-country traditions.

And yet the fact is notorious, that some of the best corn crops (I do not speak now of exceptional and premium crops), are grown in precisely this primitive way; given a good sod, and a good top-dressing turned under—with, perhaps, a little dash of superphosphate upon the hills to quicken germination, and give vigorous start,—and the New England farmer, if he give clean and thorough culture—which, under such circumstances, involves little labor—can count upon his forty or fifty bushels of sound corn to the acre. And the Scotchman or Englishman may tear the sod, or ridge the field, or drill it, or torment it as he will, before planting, and the chances are, he will reap, with the same amount of fertilizers, a smaller harvest. And it is precisely this undervaluation of his traditional mode of labor, that makes him show a distaste for the crop.

Corn is a rank grower, and, very largely, a surface feeder; for these reasons, it accommodates itself better than most farm crops, to an awkward and careless husbandry—provided only, abundance of gross fertilizers are present, and comparative cleanliness secured. It is not a crop which I should count a valuable assistant in bringing the sandy loam of a neglected farm into a condition of prime fertility. It has so rank an appetite for the inorganic riches of a soil, as to forbid any accumulation of that valuable capital. Nor do I clearly perceive how, in the neighborhood of large towns, and upon light soils, it can be made a profitable crop at the East. It has a traditional sanctity, to be sure; and a great many pleasant old gentlemen of New England, who count themselves shrewd farmers, would as soon think of abandoning their heavy ox-carts, or of adopting a long-handled shovel, as of abandoning their yearly growth of corn.

I think I have given the matter a fair test, notwithstanding the objections of my Somersetshire friend, and have added to my own experience, very much observation of my neighbors' practice. And I am very confident that if only a fair valuation be placed upon the labor and manures required, that any average corn crop grown upon light soils at the East, will cost the producer four years out of five, ten per cent. more than the market price of the Western grain. In this estimate, I make due allowance for the value of the stalks and blades for forage.

I shall enter into no array of figures for the sake of proving this point; figures can be made to prove, or seem to prove so many things. And however clearly the fact might be demonstrated, there are two classes at least, upon whom the demonstration would have no effect; the first being those over-shrewd old men, who keep unflinchingly to their accustomed ways, counting their own labor for little or nothing (in which they are not far wrong); and the other class consisting of those retired gentlemen who bring so keen a relish for farming to their work, that they rather enjoy producing a crop at a cost of twice its market value. I heartily wish I were able to participate in such pleasant triumphs.

But if the economy of maize growing for the grain product be questionable, there can be no question whatever of cultivating the crop as a forage plant, for green cutting, and for soiling purposes. In no way can a full supply of succulent food be furnished more cheaply for a herd of cows, during the heats of August and September. For this object, I have found the best results in drilling eighteen inches apart, upon inverted sod, thoroughly manured; to insure successive supplies, the sowing should be repeated at intervals of a month, from the twentieth of April to the twentieth of July. A later sowing than this last, will expose the blades to early frosts.

The amount of green food which can be cut from an acre of well-grown corn is immense; but let no one hope for successful results, without a most ample supply of manure, and clean land. The practice has fallen into disfavor with many, from the fact that they have given all their best fertilizers to other crops, and then made the experiment of growing corn-fodder with a flimsy dressing, and no care. They deserved to fail. It is to be observed moreover, that as the crop matures no seed, it makes little drain upon the mineral wealth of the land, and can be followed by any of the cereals. This suggests a simple and short rotation: First, corn—grown for its blades and stalks only (the first cuttings being succeeded by turnips): Second, carrots, Mangel, or potatoes: Third, oats or other cereal: and Fourth, clover with grass seeds, to be mown so long as the interests of the dairy or the land may demand.

A professed grain-grower, or an English farmer, would smile at such an unstudied rotation; but I name it in all confidence, as one adapted to dairy purposes, upon lands which need recuperation. It is, in fact, a succession of two fallow crops, and with proper culture and dressings, will insure accumulating fertility.

Such a simple course of green cropping is, moreover, admirably adapted to the system of soiling, which, upon all light and smooth lands, adapted to dairy purposes, in the neighborhood of towns, must sooner or later become the prevailing method; and this,—because it is economic,—because it is sure, and because it supplies fourfold more of enriching material than belongs to any other system. I am not writing a didactic book, or offering any challenge to the agricultural critics (who, I am afraid, are as full of their little jealousies as the literary critics),—else I would devote a full chapter to this theory of soiling, and press strongly what I believe to be its advantages.

The reader is spared this; but he must pardon me a little fanciful illustration of the subject, in which I have sometimes indulged, and which may, possibly, at a future day, become real.

An Illustration of Soiling.

FROM the eighty-acre flat below—so like a carpet, with its checkered growth—I order every line of division fence to be removed; the best of the material being kept in reserve for making good the border fences, and the remainder cut, split, and piled for the fire. The neighbors, who cling to the old system of two-acre lots, and pinched door-yards, open their eyes and mouths very widely at this. The novelty, like all novelties in a quiet country region, is at once astounding and oppressive. As if the parish parson were suddenly to come out in the red stockings of a cardinal, or a sober-sided select-man to appear on the highway without some important article of his dress.

I fancy two or three astute old gentlemen leaning over the border fence, as the work of demolition goes on.

"The Squire's makin' this ere farm inter a parade-ground, a'*n*'t he?" says one; and there is a little, withering, sarcastic laugh of approval.

Presently, another is charged with a reflection which he submits in this shape: "Ef a crittur breaks loose in sich a *rannge* as that, I raether guess he'll have a time on't." And there is another chirrupy laugh, and significant noddings are passed back and forth between the astute old gentlemen—as if they were mandarin images, and nodded by reason of the gravity of some concealed dead weight—(as indeed they do).

A third suggests that "there woant be no great

expense for diggin' o' post holes," which remark is so obviously sound, that it is passed by in silence.

The clearance, however, goes forward swimmingly. The new breadth which seems given to the land as the dwarfish fields disappear one after another, develops a beauty of its own. The Yellow-weeds, and withered wild-grasses, which had clung under the shelter of the fences, even with the best care, are all shorn away. The tortuous and irregular lines which the frosts had given to the reeling platoons of rails, perplex the eye no more.

Near to the centre of these opened fields is a great feeding-shed, one hundred feet by forty, its ridge high, and the roof sloping away in swift pitch on either side to lines of posts, rising eight feet only from the ground. The gables are covered in with rough material, in such shape as to leave three simple open arches at either end; the middle opening,—high and broad, so that loaded teams may pass beneath; the two flanking arches,—lower, and opening upon two ranges of stalls which sweep down on either side the building. These stalls are so disposed that the cattle are fed directly from carts passing around the exterior. Behind either range of cattle is a walk five feet broad; and between these walks,—an open space sixteen feet wide, traversing the whole length of the building, and serving at once as manure pit, and gangway for the teams which deposit from time to time their contributions of muck and turf. Midway of this central area is a covered cistern, from which, as occasion demands, the drainage of the stalls may be pumped up to drench the accumulating stock of fertilizing material.

This simple building, which serves as the summer quarters of the dairy, is picturesque in its outline; for I know no reason why economy should abjure grace, or why farm construction should be uncouth or tawdry.

A small pasture-close, with strong fencing—with gates that will not swag, and with abundance of running water, supplied from the hills, serves as an exercising ground for the cows for two hours each day. Othertimes, throughout the growing season, they belong in the open and airy stalls. The crops which are to feed them, are pushing luxuriantly within a stone's throw of their quarters. An active man with a sharp scythe, a light horse-cart and Canadian pony, will look after the feeding of a herd of fifty, with time to spare for milking and stall cleaning.

From the tenth of May to the first of June, perhaps nothing will contribute so much to a full flow of milk, as the fresh-springing grass upon some outlying pasture on the hills. After this, the cows may take up their regular summer quarters in the building I have roughly indicated. From the first to the tenth of June, there may be heavy cuttings of winter rye; from the tenth of June to the twentieth, the lucerne (than which no better soiling crop can be found) is in full season; after the twentieth, clover and orchard grass are in their best condition, and retain their succulence up to the first week in July, when, in ordinary seasons, the main reliance—maize which was sown in mid-April, is fit for the scythe. Succeeding crops of this, keep the mangers of the cows full, up to an early week in October. Afterward may come cuttings of late-sown barley, or the leaves of the Mangel, or carrot-tops, with which, as a *bonne bouche*, the cattle are withdrawn to their winter quarters, for their dietary of cut-feed, oil-cake, occasional bran and roots.

They leave behind them in their summer banqueting house, a little Rhigi of fertilizing material—not exposed to storms, neither too dry nor too moist, and of an unctuous fatness, which will make sundry surrounding fields, in the next season, carry a heavier burden than ever of purple Mangel, or of shining maize-leaves.

I perceive, too, very clearly, in furtherance of the illustration, that one acre will produce as much nutritive food, under this system, as four acres under the old plan of waste—by poaching—and by exposure of all manurial material to the fierce beat of the sun, and to the washings of rain storms. I perceive that the land, as well as cattle, are all fairly in hand, and better under control. If at any time the season, or the market, should indicate a demand for some special crop, I am not disturbed by any apprehension that this or that enclosure may be needed for grazing, and so, bar the use. I perceive that a well-regulated system must govern all the farm labor, and that there will be no place for that looseness of method, and carelessness about times and details, which is invited by the old way of turning cattle abroad to shirk for themselves.

No timid team will be thrashed, in order to wipe the fence posts with the clattering whiffletree, at the last bout around the headlands. There will be no worrying of the Buckeye in old and weedy corners; not a reed or a Golden-rod can wave anywhere in triumph. The eye sweeps over one stretch of luxuriant field, where no foot of soil is wasted. The crops, in long even lines, are marked only by the successive stages of their growth, and by their coloring. There are no crooked rows, no gores, no gatherings.

If the reader has ever chanced to sail upon a summer's day up the river Seine, he will surely remember the beautiful checker-work of crops, which shine, in lustrous green, on either bank beyond the old Norman city of Rouen. Before yet the quaint and gorgeous towers of the town have gone down in the distance, these newer beauties of the cleanly cultivated shore-land challenge his wonder and admiration. I name the scene now, because it shows a cultivation without enclosures; nothing but a traditional line—which some aged poplar, or scar on the chalk cliff marks,—between adjoining proprietors; a belt of wheat is fringed with long-bearded barley; and next, the plume-like tufts of the French trefoil, make a glowing band of crimson. A sturdy peasant woman, in wooden sabots, is gathering up a bundle of the trefoil to carry to her pet cow, under the lee of the stone cottage that nestles by the river's bank.

And I indulge my fancy with the idea of some weazen-faced New England farmer looking down upon all this from some shattered loop-hole of the wrecked chateau Gaillard, and saying—"Gosh, ef a crittur were to break loose, I guess they'd have a time on't."

There are some things we New England farmers have not learned yet.

An Old Orchard.

A CERTAIN proportion of mossy, ragged orcharding belongs to almost every New England farm. My own, in this respect, was no exception; if exceptional at all, the exception lay in the fact that its orcharding was less ragged and mossy than most; the trees were also, many of them, grafted with sorts approved twenty years ago. Eight acres of a somewhat gravelly declivity, were devoted to this growth, of which four were in apple trees, two in cherries, and two in pears. Intervals of two acres each, on either side the cherries, of unoccupied land, were in the old time planted respectively with plums and peaches. Of these, only a few ragged stumps, or fitful and black-knotted shoots, remained. Their life as well as their fruitfulness had gone by; and I only knew of them through the plaintive laments of many an old-time visitor, who tantalized me with his tales of the rare abundance of luscious stone-fruits, which once swept down the hillside.

The whole enclosure of twelve acres had relapsed into a wild condition. The turf was promiscuously an array of tussocks of wild-grass, dwarfed daisies, struggling sorrel, with here and there a mullein lifting its yellow head, and domineering over the lesser wild growth. Occasional clumps of hickory, or of wild-cherry, had shot up, and exhibited a succulence and vigor which did not belong to the cultivated trees.

And now I am going to describe fully—keeping nothing back—the manner in which I dealt with this wilderness of orchard. It was not in many respects the best way; but the record of errors in so experimental a matter, often carries as good a lesson as the record of successes. This is as true in state-craft as with old orcharding.

First, I extirpated every tree which was not a fruit tree—with the exception of one lordly sugar maple at the foot of the declivity, and standing within one of the unoccupied belts. Its stately, compact head, shading a full half acre of ground, still crowns the view. I am aware that it is an agricultural enormity. The mowers complain that the broken limbs, torn down by ice storms, are a pest; the tenant complains of its deep shade; one or two neighboring sawyers have made enticing propositions for its stalwart bole, yet I cannot forego my respect for its united age and grace.

With this exception, I made full clearance, and turned under, by careful ploughing, all the wild sod. I dressed the whole field heavily with such fertilizers as could be brought together, from home resources and from town stables, with certain addenda of lime and phosphates. I removed all trees in a dying condition, of which there were at least twenty per cent. of the gross number; I pruned away all dead limbs, all interlacing boughs, and swamps of shoots from the roots. The mosses, cocoons, and scales of old bark were carefully scraped from the trunks and larger limbs, which were then washed thoroughly with a strong solution of potash. Even at this stage of the proceedings, I felt almost repaid by the air of neatness and cleanliness which the old orchard wore; and I am sorry to say that in regard to very many of the trees, it was all the repayment I have ever received.

Among the apple trees was a large number of that old favorite, the Newtown pippin; and these, I am sorry to say, were the most mossy and dilapidated of all; nor did they improve. No scrapings or prunings tempted them to any luxuriance of growth. One by one they have been cut away, until now only two remain. The nurserymen tell us that the tree is not adapted to the soil and climate of New England. I can confirm their testimony with unction.

There was, also, a stalwart company of trees bearing that delightful little dessert fruit—the Lady apple. And I think my pains added somewhat to their thrift; they are sturdy, and full of leaves every summer; and every May, in its latter days, sees them a great pyramid of blooming and blushing white. But after the bloom, the beauty is never fully restored. There is fruit indeed, but small, pinched, pierced with curculio stings, bored through and through with the worm of the apple-moth; and over and above all, every apple is patched with a mouldy blight which forbids full growth, and gives it, with its brilliant red cheek, a falsified promise of excellence. I have found in the books no illustration of this peculiar distemper which attacks the Lady apple; but in my orchard, in the month of November, the illustrations abound.

The Esopus Spitzenberg, that red, spicy bit of apple-flesh, had its representatives among the old trees which came under my care; I may give it the credit of showing grateful cognizance of the labor bestowed. The trees thrived; they are thrifty now; the bloom is like that of a gigantic, out-spread Weigelia. The fruit too (such as the curculio spares), is full and round; but there is not a specimen of it which is not bored through by the inevitable grub of the apple-moth.

Besides the varieties I have particularized, there were the Tallman and Pound Sweetings sparsely represented; and the Rhode Island Greening, which, I will fairly admit, has made a better struggle against adverse influences, than any winter fruit I have named. So fair a struggle, indeed, that if I could only forego the visitations of the curculio and of the moth, I might hope for an old-time fulness of crop. The Strawberry apple, by reason, I think, of its early maturity (and the same is true of the Red Astrachan), has shown a more kindly recognition of care than the later fruits. The moth, if it attacks, does not destroy it. I count upon its brilliant coloring, and its piquant acidity in the first days of August, as surely as I count upon the rains which follow the in-gathering of the hay. There remained a few trees of various old-fashioned sorts, such as the Fall-Pippin, the Pearmain, the Cheseborough Russet, and the black Gilliflower, which have shown little thrift, and borne no fruit of which a modest man would be inclined to boast.

In short, there appeared so little promise of eminent results, that after two or three years I gave over all special culture of the majority of the trees, and devoting the land to grass, left them to struggle against the new sod as they best could. Fruit growers and nursery men will object that the trial was not complete; and they will, with good reason, aver that no fruit trees can make successful struggle against firmly rooted grass. From all tilled crops, within whose lines there are spaces of the brown soil subject to the dews and atmospheric influences, trees will steal the nourishment; but grass, with its serried spear-blades covering the ground, steals from the tree. An open fallow with crops in the intervals, would certainly, if sustained for a period of years, have contributed far greater thrift than the trees now possess. But an open fallow is no protection against the curculio and the apple moth. If there be a protection so simple, and of such proportions as to admit of its application to a marketable crop, I am not yet informed of it. A few worthy old gentlemen of my acquaintance, catch a few millers in a deep-necked bottle, baited with molasses, which is hung from the limbs of some favorite tree overshadowing their pig-pen; and they point with pride to the results. I certainly admire their successes, but have not been tempted to emulate them, on the extended scale which the mossy orchard would have afforded.

Some persistent amateurs and pains-taking gentlemen do, I know, succeed in making the young fruit of a few favorite plum trees distasteful to the curculio, by repeated ejections of a foul mixture of tobacco and whale-oil soap,—by which the tree has a weekly bath, and an odor of uncleanness. But in view of a large orchard, where apples make a leafy pyramid measured by cubic yards, and cherries carry their fine fruit sixty feet in the air, there would be needed a projectile of dirty water that would rival Alderman Mechi's, of Tip-tree Hall.

It is far easier to accomplish successful results with an old orchard of native, wild growth, than with one of grafted fruit;—even as the Doctors find that a reprobate who has fallen away from grace and early good conduct, is a worse subject for reformation, than an unkempt savage.

The grafted tree wants an abounding luxuriance of material, from which to elaborate its exceeding size and flavor; and if by neglect, this material be wanting, the organs of its wonderful living laboratory shrink—from inaction, and part with a share of their vitality. The native tree, on the other hand, having no special call upon it for the elaboration of daintier juices than go to supply a cider vat, has steady normal development under all its mosses, and retains a stock of reserved vitality, which, if you humor with good tillage and dressings, and point with good grafts, will carry a good tale to the apple bin.

On the very orchard I have named, were some two or three uncouth, lumbering, unpromising trees, yet sound as a nut to their outermost twigs, which the simple dressings, tillage, and washings that were bestowed somewhat vainly upon the others, quickened into a marvellous luxuriance; and the few shoots I set upon them are now supplying the best fruit of the orchard. Even these, however, are not free from the pestilent stings which the swarms of winged visitors inflict upon every crop.

It is very questionable if ploughing is, upon the whole, the best way of reinstating a neglected and barren orchard. It is a harsh method; trees struggling to keep up a good appearance under adverse circumstances—like men—use every imaginable shift; their little spongiole feeders go off on wide search; they are multiplied by the diversity of labor; and the plough cuts into them cruelly, making crude butcher work where the nicest surgery is demanded. I am inclined to believe that a deep trench, sunk around each tree, at the distance of from eight to ten feet from the trunk, and filled with good lime compost, is the surest way of redeeming a neglected orchard. Even then, however, the turf should be carefully removed within the enclosed circle, that the air and its influences may have penetrative power upon the soil. The method is Baconian (*fodiendo et aperiendo terram circa radices ipsarum*); it is thorough, but it is expensive; and a farmer must consider well—if his trees, soil, market, and the populousness of the insect world will warrant it.

For my own part, so far as regards a market crop

of winter fruit, I have decided very thoroughly in the negative. Not that it cannot be grown with sufficient care; but that it can be grown far more cheaply, and of a better quality, in other regions. Summer fruit is not so long exposed to the depredations of insects, nor will it bear distant transportation. Its freshness too, gives it a virtue, and a relishy smack, which warrant special pains-taking.

I find in an old book of Gervase Markham's, "The Countrie Farme" (based upon Liebault), that the apple tree "loveth to have the inward part of his wood moist and sweatie, so you must give him his lodging in a fat, black, and moist ground; and if it be planted in a gravelly and sandie ground, it must be helped with watering, and batling with dung and smal moulde in the time of Autumne. It liveth and continueth in all desirable good estate in the hills and mountains where it may have fresh moisture, being the thing that it searcheth after, but even there it must stand in the open face of the South."

The ruling is good now, with the exception perhaps of exposure to the South, in regions liable to late spring frosts. And whatever may be the advantages of soil and of position, let no man hope for large commercial results in apple-growing at the East, without reckoning upon as thorough and assiduous culture as he would give to his corn crop;—

as well as a constant battle with the borers and bark lice,—intermittent campaigns against the caterpillar and canker worm, and a great June raid upon the whole guerilla band of curculios.

The cherries, a venerable company of trees, have borne the scrapings and dressings with great equanimity,—being too old to be pushed into any wanton luxuriance, and too sedate to show any great exhilaration from the ammoniacal salts. Pruning is not much recommended in the books; yet I have succeeded in restoring a good rounded head of fruit-bearing wood by severe amputation of begummed and black-stained limbs; this is specially true of the Black-hearts and Tartarians,—of many of which I have made mere pollards.

It is a delicate fruit to be counted among farm crops, and hands used to the plough are apt to grapple it too harshly. Pliny says it should be eaten fresh from the tree; and it is as true of our best varieties, as it was of the Julian cherry in the first century. It will not tolerate long jogging in a country wagon; it will not "keep over" for a market; and between these drawbacks, and the birds—who troop in flocks to the June feast,—and the boy pickers—who take toll as they climb,—and the outstanding twigs, which shake defiance to all ladders and climbers—I think he is a fortunate man who can market from forty-year-old trees, one bushel in three.

Of the position for a cherry orchard, and of its likings in the way of soil and climate, nothing better can be said, than Palladius wrote fourteen centuries ago: "*Cerasus amat cœli statum frigidum, solum vero positionis humectæ. In tepidis regionibus parva provenit. Calidum non potest sustinere. Montana, vel in collibus constituta regione lætatur*," *—which means that—cherries want a cool air and moist land. Heat hurts them, and makes them small, and they delight in a hilly country.

The Pears.

THE condition of the pears was far worse than that of either cherries or apples. Had they been seedlings of the native fruit, they would have shown more stalwart size, and better promise from good treatment. There was, I remember, a long weakly row of the Madeleine, shrouded in lichens, and with their lank, frail limbs all tipped with dead wood. It is an enticing fruit, by reason of its early ripening, and its pleasant sprightly flavor; but its persistent inclination to rot at the core, in most soils, makes it a very unprofitable one. I forthwith cut

* Lib. xi., Tit. 12.

away their dying, straggling tops, and by repeated diggings about the roots, stimulated a growth of new wood, upon which luxuriant grafts are now (six years after commencement of operations) bearing full crops of more approved varieties. The Jargonelles were almost past cure. Long struggle with neglect had nearly paralyzed their vegetative power; but by setting a few scions of such rank growers as the Buffum upon the most promising of the purple shoots, I have met with fair success. The Jargonelle itself, I may remark in passing, seems to me not fitly appreciated in the race after new French varieties. It has a juiciness, a crispness, and a vinous flavor, which, however scorned by the later pomologists, are exceedingly grateful on a hot August day.

There was a great rank of Virgouleuse (white Doyenne)—pinched in their foliage, with bark knotted like that of forest trees, and bearing only cracked, meagre, woody fruit. For New England it is a lost variety. Happily, however, its boughs take grafts with great kindliness; and I have now the pleasure of seeing fair full heads upon every one of these out-lived stocks, of the Bartlett, Flemish Beauty, Bonne de Jersey, and Lawrence.

There were not a few Buffum trees in the ranks, which were in a state of most extraordinary dilapidation; their trunks white with moss, their upright shoots completely covered with a succession of crooked, gnarled, mossy fruit spurs, that crinkled under the scraper like dried brambles; the extremity of every upright bough was reduced to a shrivelled point of blackened and sun-dried wood, and the fruit so dwarfed as to puzzle the most astute of the pomologists.

I made a clean sweep of the old fruit spurs, —docked the limbs,—scraped the bark to the quick, —washed with an unctuous soapy mixture,—dug about and enriched the roots, and in three years' time, there were new leading shoots, all garnished with fresh fruit spurs—which, in September fairly broke away with the weight of the glowing pears.

The Seckels, of which there were several trees, have not come so promptly 'to time.' The fertilizers and the cleaning process, which have given rampant vigor to the Buffums, have scarce lent to the dwindled Seckels any appreciable increase of size or of succulence. The same is true, in a less degree, of certain old stocks, grafted some fifteen years ago with Bonne de Jersey, and since left to struggle with choking mosses, and wild sod.

It is unnecessary to enumerate all the varieties which I found stifling in my orchard,—from the bright little Harvest pear to the crimson-cheeked

Bon-Chrétien. Here and there I have religiously guarded some old variety of Sugar-pear, or of Bergamot,—by reason of the pleasant associations of their names, and by reason of an old fashioned regard which I still entertain for their homeliness of flavor. I sometimes have a visit from a pear-fancier, who boasts of his fifty or hundred varieties,—who confounds me with his talk of a Beurre St. Nicholas, or a Beurre of Waterloo, and a Doyenne Goubault, or a Doyenne Robin; I try to listen, as if I appreciated his learning; but I do not. My tastes are simple in this direction; and I feel a blush of conscious humility when he comes upon one of my old-time trees, staggering under a load of fruit—which are not in the books. It is very much as if a gentleman of the Universities, full of his book lore, were to stroll into my library,—talking of his Dibdins, and Elzevirs, and Brunets;—with what a blush I should see his eye fall upon certain thumb-worn copies of Tom Jones, or the Vicar of Wakefield, or Defoe!

Yet these gentlemen of the special knowledges have their uses—the pear-mongers with the rest. Not a season passes, but they discover and label for us a host of worthless varieties. I only object to the scornful way in which they ignore a great many established favorites, which people *will* persist in buying and eating. I remember that I once had the hardihood, in a little group of pomological gentlemen, to express a modest opinion in praise of the flavor of the Bartlett pear.

The gentlemen did not deign a reply; but I was looked upon very much as a greenhorn might be, who at a political caucus should venture a word or two, in favor of—honesty.

Quince stocks for pear trees have their advocates; and there has been a very pretty war between the battlers for the standards, and the battlers for the dwarfs. Having made trial of both, and considering that most human opinions are fallible, I plant myself upon neutral ground, and venture to affirm that each mode of culture has its advantages. There are, for instance, varieties of the pear, which, in certain localities, will not thrive, or produce fair specimens, without incorporation upon the quince stock. Such, in my experience, are the Duchess d'Angoulême, and the Vicar of Winkfield. The finest fruit of the Belle Lucrative, and the Bonne de Jersey, I also invariably take from dwarf growth.

The dwarf trees, however, demand very special and thorough culture; if the season is dry, they must be watered; if the ground is baked, it must be stirred. I look upon them as garden pets, which must be fondled and humored; and like other pets, they are sure to be attacked by noxious diseases.

7*

They take the leaf-blight as easily as a child takes the mumps; they are capricious and uncertain—sometimes repaying you for your care well; and other times, dropping all their fruit in a green state, in the most petulant way imaginable. And worst of all, after two or three years of devoted nursing, without special cause, and with all their leaves laughing on them, some group of two or three together—suddenly die.

Early bearing, and brilliant specimens favor the quince; but hardiness, long life, and full crops favor the pear upon its own roots. If a man plant the latter, he must needs wait for the fruit. Mœris puts it very prettily in the Eclogue:—

"Insere, Daphni, *pyros:* carpent tua poma nepotes."

But if a man with only a few perches of garden, and with an aptitude for nursing, desires fruit the second or third year after planting, let him by all means —plant the dwarfs. Yet even then his success is uncertain,—particularly if he indulges in the "latest varieties." I am compelled to say that I have known several cautious old gentlemen, who—with a garden full of dwarf trees,—have been seen in the month of September, to slip into a fruit shop at the edge of evening, with suspicious-looking, limp panniers on their arms. Nay,—I have myself met them returning from such furtive errand, with a basket laden from the fruiterer's stock, carefully hidden under their skirts; and I have gone my way—(pretending not to see it all), humming to myself,

—— carpent tua poma nepotes!

Want of success in orcharding is more often attributable to want of care, than to any other want whatever. There are, indeed, particular belts of land which seem to favor the apple,—where, with only moderate cultivation, they are free from leaf blight,—comparatively free from insect depredators, and fruit with certainty. There are other regions, —and these, so far as I have observed, warm soils inclining to a sandy or gravelly loam, in which the apple does not show vigor, except under extraordinary attention, and in which the whole insect tribe seems doubly pestiferous.

The pear is by no means so capricious; it will thrive in a heavy loam; it will thrive in light sand; the borer does not attack its root; the caterpillar moth does not fasten its eggs (or very rarely) upon its twigs; the apple-moth spares a large proportion of its fruit. But even the pear, without care and cultivation, will disappoint; and the farmer who neglects any crop, will find, sooner or later, that whatever is worth planting, is worth planting well; whatever is worth cultivating, is worth cultivating well; and that nothing is worth harvesting, that is not worth harvesting with care.

My Garden.

I ENTER upon my garden by a little, crazy, rustic wicket, over which a Virginia creeper has tossed itself into a careless tangle of festoons. The entrance is overshadowed by a cherry-tree, which must be nearly half a century old, and which, as it filches easily very much of the fertilizing material that is bestowed upon the garden, makes a weightier show of fruit than can be boasted by any of the orchard company.

A broad walk leads down the middle of the garden,—bordered on either side by a range of stout box, and interrupted midway of its length by a box-edged circle, that is filled and crowned with one cone-shaped Norway-Spruce. These lines, and this circlet of idle green, are its only ornamentation. Easterly of the walk is a sudden terrace slope, stocked with currants, raspberries, and all the lesser fruits, in a maze of lines and curves. Westward is a level open space, devoted to long parallel lines of garden vegetables. The slope, by reason of its surface and its crops, is subject only to fork-culture; the western half, on the other hand, has the economy of deep and thorough trench-ploughing, every autumn and spring.

Nor is this an economy to be overlooked by a farmer. Very many, without pretensions to that nicety of culture which is supposed to belong to spade husbandry alone, so overstock their gardens with confused and intercepting lines of fruit shrubbery, and perennial herbs, as to forbid any thorough action of the plough. By the simple device, however, of giving to the garden the shape of a long parallelogram, and arranging its trees, shrubbery, and walks, in lines parallel with its length, and by establishing easy modes of ingress and egress at either end, the plough will prove a great economizer; and under careful handling, will leave as even a surface, and as fine a tilth as follows the spade. I make this suggestion in the interest of those farmers who are compelled to measure narrowly the cost of tillage, and who cannot indulge in the amateur weakness of wasted labor.

I have provided also a leafy protection for this garden against the sweep of winds from the northwest: northward, this protection consists of a wild belt of tangled growth—sumacs, hickories, cedars, wild-cherries, oaks—separated from the northern walk of the garden, by a trim hedge-row of hemlock-spruce. This tangled belt is of a spontaneous growth, and has shot up upon a strip of the neglected pasture-land, from which, seven years since, I trenched the area of the garden. Thus it is not only a protection, but offers a pleasant contrast of what the whole field might have been, with what the garden now is. I must confess that I love these savage waymarks of progressive tillage—as I love to meet here and there, some stolid old-time thinker, whom the rush of modern ideas has left in picturesque isolation.

Time and again some enterprising gardener has begged the privilege of uprooting this strip of wildness, and trenching to the skirt of the wall beyond it; but I have guarded the waste as if it were a crop; the cheewits and thrushes make their nests undisturbed there. The long, firm gravel-alley which traverses the garden from north to south, traverses also this bit of savage shrubbery, and by a latticed gate, opens upon smooth grass-lands beyond, which are skirted with forest.

Within this tangle-wood, I have set a few graftlings upon a wild-crab, and planted a peach or two—only to watch the struggle which these artificial people will make with their wild neighbors. And so various is the growth within this limited belt, that my children pick there, in their seasons,—luscious dew-berries, huckleberries, wild raspberries, bill-berries, and choke-cherries; and in autumn, gather bouquets of Golden-rod and Asters, set off with crimson tufts of Sumac, and the scarlet of maple boughs. And when I see the brilliancy of these, and smack the delicate flavor of the wild-fruit, it makes me doubt if our progress is, after all, as grand as it should be, or as we vainly believe it to be; and (to renew my parallel)—it seems to me that the old-time and gone-by thinkers may possibly have given us as piquant, and marrowy suggestions upon whatever subject of human knowledge they touched, as the hot-house philosophers of to-day. I never open, of a Sunday afternoon, upon the yellowed pages of Jeremy Taylor, but his flavor and affluence, and homely wealth of allusions, suggest the tangled wild of the garden—with its starry flowers, its piquant berries, its scorn of human rulings, its unkempt vigor, its boughs and tendrils stretching heaven-ward; and I never water a reluctant hill of yellowed cucumbers, and coax it with all manner of concentrated fertilizers into bearing,—but I think of the elegant education of the dapper Dr. ———, and of the sappy, and flavorless results.

To the westward of the garden, and concealing a decrepit mossy wall, that is covered with blackberry vines and creepers, is the flanking shelter of another hemlock hedge of wanton luxuriance. A city garden could never yield the breadth it demands; but upon the farm, the complete and graceful protection it gives, is well purchased, at the cost of a few feet of land. Nor is much time required for its growth; five years since, and this hedge of four feet in height, by two hundred yards in length, was all brought away from the wood in a couple of market baskets.

The importance of garden shelter is by no means enough considered. I do not indeed name my own method as the best to be pursued; flanking buildings or high enclosures may give it more conveniently in many situations; a steep, sudden hillside may give it best of all; but it should never be forgotten that while we humor the garden soil with what the plants and trees best love, we should also give their foliage the protection against storms which they covet; and which, in an almost equal degree, contributes to their luxuriance.

To the dwarf fruit, as well as to the grape, this shelter is absolutely essential; if they are compelled to fortify against aggressive blasts,—they may do it indeed; but they will, in this way, dissipate a large share of the vitality which would else go to the fruit. Young cattle may bear the exposure of winter, but they will be pinched under it, and take on a weazen look of age, and expend a great stock of vital energy in the contest.

Fine Tilth makes Fine Crops.

WITH a good situation, the secret of success with garden crops, lies in the richness of the soil, and in its deep and fine tilth; the last being far oftener wanting than the former. A farm crop of potatoes or even of corn, will make a brave struggle amid coarse nuggets of earth, if only fertilizers are present; but such fine feeders as belong to the garden can lay no hold upon them; they want delicate diet. Farmers are often amazed by the extraordinary vegetable results upon the sandy soil of a city dooryard, which they would count comparatively worthless; not considering, that—aside from the shelter of brick walls, which make the sun do double duty—the productive capacity of such city gardens, lies very much in the extreme and almost perfect comminution of the soil.

What is true of garden earth, is true also of its fertilizers; they must be triturated, fine, easily digestible. Masses of unbroken farm-yard material are no more suited to the delicate organization of garden-plants, than a roasted side of bacon is suited to a child's diet. They may struggle with it indeed.

Possibly they may reduce it to subjection; but their growth will be rank and flavorless, whatever size they may gain.

It is a common mistake to suppose that garden products are good in proportion to their size. The horticultural societies have done great harm in bolstering the admiration for mere grossness. Smoothness, roundness, perfect development of all the parts, and delicacy of flavor, are the true tests. I remember once offering for exhibition a little tray of garden products, in which every fruit and vegetable—though by no means all they should have been—was perfect in outline, well developed, free from every sting of insect or excrescence, and of that delicate and tender fibre which belongs only to swift and unchecked growth; yet my poor tray was overslaughed entirely, by an adjoining show of monster vegetables, with warty excrescences, and of rank and wholly abnormal development. The committee would have been properly punished if they had been compelled to eat them.

In the same way, and with equal fatuity, the societies for agricultural encouragement persist in giving premiums to—so called—fat cattle; mere monsters—not of good, wholesome, muscular fibre, well-mottled—but mountains of adipose substance, which no Christian can eat, and which are only disposed of profitably, by serving as an advertisement to some venturesome landlord, from whose table the reeking fat goes to the soap-pot.

Grossness does not absorb excellence, or even imply it—either in the animal or vegetable world. I have never yet chanced to taste the monstrosities which the generous Californians sometimes send us in the shape of pears; but without knowing, I would venture the wager of a bushel of Bartletts, that one of our own, little, jolly, red-cheeked Seckels would outmatch them thoroughly—in flavor, in piquancy, and in vinous richness.

Shall the flaunting Dahlia match us a Rose? Yet the dahlia has its place too; it gives scenic effect; its tall stiffness tells in the distance; but we have a thousand roses at every hand.

I sometimes fear that this disposition to set the mere grossness of a thing above its finer qualities, is an American weakness. We do not forget, so often as we might to advantage—that we are a great people. That eagle which our Fourth of July orators paint for our delighted optics, dipping his wings in both oceans, is the merest buzzard of a bird, except he have more virtue in him than mere size.

Seeding and Trenching.

IF there is one fault above another in all the gardening books, it is the lack of those simplest of directions and suggestions, without which the novice is utterly at fault. Thus, we are told in what month to sow a particular seed—that it must have a loamy soil; and are favored with some special learning in regard to its varieties, and its Linnæan classification.

"Pat," we say, "this seed must be planted in a loamy soil."

Pat, (scratching his head reflectively); "And shure, isn't it in the garden thin, ye'd be afther planting the seed?"

Pat's observation is a just one; of course we buy our seed to plant in the garden, no matter what soil it may love. The more important information in regard to the depth of sowing it, the mode of applying any needed dressing, the requisite thinning, the insect depredators, and the mode of defeating them—is, for the most part, withheld. That the matter is not without importance, one will understand who finds, year after year, his more delicate seeds failing, and the wild and attentive Irishman declaring,—

"And, begorra thin, it's the ould seed."

"But did you sow it properly, Patrick?"

"Didn't I, faith? I byried 'em an inch if I byried 'em at all."

An inch of earth will do for some seeds, but for others, it is an Irish burial—without the wake.

The conditions of germination are heat, air, and moisture. Covering should not be so shallow as to forego the last, nor so deep as to sacrifice the other essential influences. Heat alone will not do; air and moisture alone will not do. A careful gardener will be guided by the condition of his soil, and the character of his seed. If this have hard woody covering like the beet, he will understand that it demands considerable depth, to secure the moisture requisite to swell the kernel; or that it should be aided by a steep, before sowing. If, on the other hand, it be a light fleecy seed, like the parsnip, he will perceive the necessity of bringing the earth firmly in contact with it.

As a general rule, the depth of covering should not exceed two or three times the shortest diameter of the seed; this plainly involves so light a covering for the lettuces, parsley, and celery, that a judicious gardener will cover by simply sifting over them a sprinkling of fine loam, which he will presently wet down thoroughly (unless the sun is at high noon), with his water-pot—medicined with a slight pinch of guano.

For a good garden, as I have said, a deep rich soil is essential; and to this end trenching is desirable; but trenching will not always secure it, for the palpable reason that subsoil is not soil. I have met with certain, awkward confirmatory experiences,—where a delicate garden mould of some ten inches in depth, which would have made fair show of the lesser vegetables, has been, by the frenzy of trenching, buried under fourteen inches of villainous gravelly hard-pan, brought up from below, in which all seeds sickened, and all plants turned pale. Whatever be the depth of tillage, it is essential that the surface show a fine tilth of friable, light, unctuous mould; the young plants need it to gain strength for a foray below. And yet I have seen inordinate sums expended, for the sake of burying a few inches of such choice moulds, under a foot-thick coverlid of the dreariest and rawest yellow gravel that ever held its cheerless face to the sun.

The amateur farmer, however, is not staggered by any such difficulties; indeed, he courts them, and delights in making conquest. They make good seed-bed for his theories—far better than for his carrots. Let me do no discredit, however, to 'trenching,' which in the right place, and rightly performed, by thorough admixture, is most effective and judicious; nor should any thoroughly good garden be established upon soil which will not admit of it, and justify it. If otherwise, my advice is, not to trench, but—sell to an amateur.

How a Garden should Look.

THE æsthetic element does not abound in the minds of country farmers; and there is not one in a thousand who has any conception of a garden, save as a patch (always weedy) where the goodwife can pluck a few condiments for dinner. If you visit one, he may possibly take you to see a 'likely yearling,' or a corn crop, but rarely to his garden. Yet there is no economic reason why a farmer's garden should not make as good and as orderly a show, as his field crops.

A straight line is not greatly more difficult to make than a crooked one. The absurd borders, indeed, where dirt is thrown into line, and beaten with a spade, is a mere caprice, which there is no need to imitate; but the neatness which belongs to true lines of plants, regular intervals between crops, perfect cleanliness, is another matter; and is so feasible and so telling in effect, that no farmer has good excuse for neglecting it. Effective groupings, again, of dwarf trees and fruit shrubbery, whether in rows, curves, or by gradations of size, give points of interest, and contribute to the attractions of a garden.

It is not a little odd that the back-country gentleman, who replies to all such suggestions, that he cares nothing for appearances—shall yet never venture to a militia muster, or a town meeting, without slipping into the 'press' for the old black-coat, and the black beaver (giving it a coquettish wipe with his elbow)—to say nothing of the startling shirt-collars, whose poise he studies before the keeping-room mirror.

He contracts too for a staring white coat of paint upon his house and palings, and a mahogany-colored door, out of the same irresistible regard to "what people will say." But in all this, he does not do one half so much for the education of his children into a perception of order and elegance, as if he bestowed the same care upon the neatness of his yard and garden, where their little feet wander every day.

It would be hard to estimate the educating effect of the gardens of the Tuileries and Luxembourg upon the minds of those artisans of Paris, who, living in garrets, and too poor for anything more than a little rustic tray of flowers upon their window ledge, are yet possessed of a perception of grace, which shines in all their handiwork. And if you transport them to the country—their own Auvergne or Normandy—they cannot, if they would, make slatternly gardens: they will not indeed repeat the brilliant tints of Paris flowers; they cannot rival the variety; but they can stamp lines of grace, and harmony of arrangement upon the merest door-yard of vegetables and pot-herbs.

Here let me outline, in brief, what a farmer's garden may be made, without other than home-labor. A broad walk shall run down the middle of either a square enclosure, or long parallelogram. A box edging upon either side is of little cost, and contributes eminently to neatness; it will hold good for eight years, without too great encroachment, and at that time, will sell to the nurserymen for more than enough to pay the cost of resetting. On either side of this walk, in a border of six feet wide, the farmer may plant his dwarf-fruit, with grapes at intervals to climb upon a home-made cedar trellis, that shall overarch and embower the walk. If he love an evening pipe in his garden, he may plant some simple seat under one or more of these leafy arbors.

At least one half the garden, as I before suggested, he may easily arrange, to till,—spring and autumn,—with the plough; and whatever he places there in the way of tree and shrub, must be in lines parallel with the walk. On the other half, he will be subjected to no such limitations; there, he will establish his perennials—his asparagus, his thyme, his sage, and parsley; his rhubarb, his gooseberries, strawberries, and raspberries; and in an angle—hidden if he choose by a belt of shrubbery—he may have his hotbed and compost heap. Fork-culture, which all these crops demand, will admit of any arrangement he may prefer, and he may enliven the groupings, and win the goodwife's favor, by here and there a little circlet of such old-fashioned flowers as tulips—yellow lilies and white, with roses of all shades.

Upon the other half he may make distribution of parts, by banding the various crops with border lines of China or Refugee beans; and he may split the whole crosswise, by a walk overarched with climbing Limas, or the London Horticultural—setting off the two ends with an abutment of Scarlet-runners, and a surbase of fiery Nasturtium.

There are also available and pretty devices for making the land do double duty. The border lines of China-beans, which will be ripened in early August, may have Swedes sown in their shadow in the first days of July, so that when the Chinas have fulfilled their mission, there shall be a new line of purple green in their place. The early radishes and salads may have their little circlets of cucumber pits, no way interfering with the first, and covering the ground when the first are done. The early Bassano beets will come away in time to leave space for the full flow of the melons that have been planted at intervals among them. The cauliflower will find grateful shade under the lines of sweet corn, and the newly-set winter cabbages, a temporary refuge from the sun, under shelter of the ripened peas. I do not make these suggestions at random, but as the results of actual and successful experience.

With such simple and orderly arrangement, involving no excessive labor, I think every farmer and country-liver may take pleasure in his garden as an object of beauty;—making of it a little farm in miniature, with its coppices of dwarf-trees, its hedge-rows of currants and gooseberries, and its meadows of strawberries and thyme. From the very day on which, in spring, he sees the first, faint, upheaving, tufted lines of green from his Dan-O'Rourkes, to the day when the dangling Limas, and sprawling, bloody tomatoes are smitten by the frost, it offers a field of constant progress, and of successive triumphs. Line by line, and company by company, the army of green things take position; the little flowery banners are flung to the wind; and lo! presently every soldier of them all—plundering only the earth and the sunshine—is loaded with booty.

The Lesser Fruits.

FROM the time when I read of Mistress Doctor Primrose's gooseberry wine, which the Doctor celebrates in his charming autobiography, I have entertained a kindly regard for that fruit. But my efforts to grow it successfully have been sadly baffled. The English climate alone, I think, will bring it to perfection. I know not how many ventures I have made with 'Roaring-Lion,' 'Brown Bob,' 'Conquerors,' and other stupendous varieties; but without infinite care, after the first crop— the mildew will catch and taint them. Our native varieties,—such, for instance, as the Houghton-seedling, make a better show, and with ordinary care, can be fruited well for a succession of seasons. But it is not, after all, the stanch old English berry, which pants for the fat English gardens, for the scent of hawthorn, and for the lowering fog-banks of Lancashire.

Garden associations (with those who entertain them) inevitably have English coloring. Is it strange—when so many old gardens are blooming through so many old books we know?

No fruit is so thoroughly English in its associations; and I never see a plump Roaring-Lion, but I think of a burly John Bull, with waistcoat strained over him like the bursting skin of his gooseberry, and muttering defiance to all the world. There is, too, another point of resemblance; the fruit is liable to take the mildew when removed from British soil, just as John gets the blues, and wraps himself in a veil of his own foggy humors, whenever he goes abroad. My experience suggests that this capricious fruit be planted under the shadow of a north wall, in soil compact and deep; it should be thoroughly enriched, pruned severely, watered abundantly, and mulched (if possible) with kelp, fresh from the sea shore. These conditions and appliances may give a clean cheek, even to the Conquering-Hero.

But it is not so much for any piquancy of flavor that I prize the fruit, as because its English bloat is pleasantly suggestive of little tartlets (smothered in clotted cream) eaten long ago under the lee of Dartmoor hills—of Lancashire gardens, where prize berries reposed on little scaffoldings, or swam in porcelain saucers—and of bristling thickets in Cowper's 'Wilderness' by Olney.

Is it lonely in my garden of a summer's evening? Have the little pattering feet gone their ways—to bed? Then I people the gooseberry alley with old Doctor Primrose, and his daughters Sophia and Olivia; Squire Burchell comes, and sits upon the bench with me under the arbor, as I smoke my pipe. How shall we measure our indebtedness to such pleasant books, that people our solitude so many years after they are written! Oliver Goldsmith, I thank you! Crown-Bob, I thank you.

Gooseberries, like the English, are rather indigestible.

Of strawberries, I shall not speak as a committee-man, but as a simple lover of a luscious dish. I am not learned in kinds; and have even had the *niaiserie* in the presence of cultivators, to confound Crimson Cone with Boston-Pine; and have blushed to my eyelids, when called upon to name the British-Queen in a little collection of only four mammoth varieties. With strawberries, as with people, I believe in old friends. The early Scarlet, if a little piquant, is good for the first pickings; and the Hovey, with a neighbor bed of Pines, or McAvoy, and Black Prince, if you please, give good flavor, and a well-rounded dish. The spicy Alpines should bring up the rear; and as they send out but few runners, are admirably adapted for borders. The Wilson is a great bearer, and a fine berry; but with the tweak of its acidity in my mouth, I can give its flavor no commendation. Supposing the land to be in good vegetable-bearing condition, and deeply dug, I know no dressing which will so delight the strawberry, as a heavy coat of dark forest-mould. They are the children of the wilderness, force them as we will; and their little fibrous rootlets never forget their longing for the dark, unctuous odor of mouldering forest leaves.

Three great traveller's dishes of strawberries are in my mind.

The first was at an inn in the quaint Dutch town of Broek: I can see now the heaped dish of mammoth crimson berries,—the mug of luscious cream standing sentry,—the round red cheese upon its platter,—the tidy hostess, with arms akimbo, looking proudly on it all: the leaves flutter idly at the latticed window, through which I see wide stretches of level meadow,—broad-armed windmills flapping their sails leisurely,—cattle lying in lazy groups under the shade of scattered trees; and there is no sound to break the June stillness, except the buzzing of the bees that are feeding upon the blossoms of the linden which overhangs the inn.

I thought I had never eaten finer berries than the Dutch berries.

The second dish was at the Douglas-Hotel in the city of Edinboro'; a most respectable British tavern, with a heavy solid sideboard in its parlor; heavy solid silver upon its table; heavy and solid chairs with cushions of shining mohair; a heavy and solid figure of a landlord; and heavy and solid figures in the reckoning.

The berries were magnificent; served upon quaint old India-china, with stems upon them, and to be eaten as one might eat a fig, with successive bites, and successive dips in the sugar. The Scotch fruit was acid, I must admit, but the size was monumental. I wonder if the stout landlord is living yet, and if the little pony that whisked me away to Salisbury crag, is still nibbling his vetches in the meadow by Holyrood?

The third dish was in Switzerland, in the month of October. I had crossed that day the Scheideck from Meyringen, had threaded the valley of Grindelwald, and had just accomplished the first lift of the Wengern Alp—tired and thirsty—when a little peasant girl appeared with a tray of blue saucers, brimming with Alpine berries—so sweet, so musky, so remembered, that I never eat one now but the great valley of Grindelwald, with its sapphire show of glaciers, its guardian peaks, and its low meadows flashing green, is rolled out before me like a map.

In those old days when we school-boys were admitted to the garden of the head-master twice in a season—only twice—to eat our fill of currants (his maid having gathered a stock for jellies two days before), I thought it 'most-a-splendid' fruit; but I think far less of it now. My bushes are burdened with both white and red clusters, but the spurs are somewhat mossy, and the boughs have a straggling, dejected air. With a little care, severe pruning, due enrichment, and a proper regard to varieties

(Cherry and White-Grape being the best), it may be brought to make a very pretty show as a dessert fruit. But as I never knew it to be eaten very freely at dessert, however finely it might look, I have not thought it worth while to push its proportions for a mere show upon the exhibition tables. The amateurs would smile at those I have; but I console myself with reflecting that they smile at a great deal of goodness which is not their own. They are full of conceit—I say it charitably. I like to upset their proprieties.

There was one of them, an excellent fellow (if he had not been pomologically starched and jaundiced), who paid me a visit in my garden not long ago, bringing his little son, who had been educated strictly in the belief that all fine fruit was made—not to be enjoyed, but for pomological consideration. The dilettante papa was tip-toeing along with a look of serene and well-bred contempt for my mildewed gooseberries and scrawny currants, when I broke off a brave bough loaded with Tartarian cherries, and handed it to the lad, with—"Here, Harry, my boy,—we farmers grow these things to eat!"

What a grateful look of wonderment in his clear gray eyes!

The broken limb, the heresy of the action, the suddenness of it all, were too much for my fine friend.

I do not think that for an hour he recovered from the shock to his sensibilities.

Of raspberries, commend me to the Red-Antwerp, and the Brinckle's Orange; but to insure good fruitage, they should be protected from high winds, and should be lightly buried, or thoroughly 'strawed over' in winter. The Perpetual, I have found a perpetual nuisance.

The New-Rochelle or Lawton blackberry has been despitefully spoken of by many; first, because the market-fruit is generally bad, being plucked before it is fully ripened; and next, because in rich clayey grounds, the briers, unless severely cut back, and again back, grow into a tangled, unapproachable forest, with all the juices exhausted in wood. But upon a soil moderately rich, a little gravelly and warm, protected from wind, served with occasional top-dressings and good hoeings, the Lawton brier bears magnificent burdens.

Even then, if you would enjoy the richness of the fruit, you must not be hasty to pluck it. When the children say with a shout,—"The blackberries are ripe!" I know they are black only, and I can wait.

When the children report—"The birds are eating the berries," I know I can still wait. But when they say—"The bees are on the berries," I know they are at full ripeness.

Then, with baskets we sally out; I taking the middle rank, and the children the outer spray of boughs. Even now we gather those only which drop at the touch; these, in a brimming saucer, with golden Alderney cream, and a *soupçon* of powdered sugar, are Olympian nectar; they melt before the tongue can measure their full roundness, and seem to be mere bloated bubbles of forest honey.

There is a scratch here and there, which calls from the children a half-scream; but a big berry on the lip cures the smart; and for myself, if the thorns draggle me, I rather fancy the rough caresses, and repeat with the garden poet (humming it half aloud):

Bind me, ye woodbines, in your twines;
Curl me about, ye gadding vines;
And oh! so close your circles lace,
That I may never leave this place;
But, lest your fetters prove too weak,
Ere I your silken bondage break,
Do you, O brambles, chain me too,
And, courteous briers, nail me through.

Grapes.

IF the associations of the gooseberry are British, those of the vine are thoroughly Judæan. There is not a fruit that we grow, which has so venerable and so stately a history. Who does not remember the old Biblical picture in all the primers, of the stupendous cluster which the spies brought away from the brook Eshcol? And I am afraid that many a youngster, comparing it with the milder growth which capped his dessert, has viewed it with a little of the Bishop-Colenso scepticism.

Upon a certain day I give to my boy,—who has worked some mischief,—the smallest bunch of the dish. He poises it in his hand awhile, looking askance—doubtful if he will fling it down in a pet, or enjoy even so little. The latter feeling wins upon him, but is spiced with a bit of satire, that relieves itself in this way:

"I think, papa (he is fresh from "Line upon Line"), that the spies wouldn't put a staff on their shoulders to carry such a bunch as *that!*"

By this admeasurement, indeed, no portion of New England can be counted equal to the land of Canaan. There are grapes, however, which yield gracefully to the requisitions of the climate, and furnish abundant clusters, if not large ones. As yet, for out-of-door culture—such as every farmer may plant with faith, and without trembling for the early frosts—the two most desirable are the Concord and Diana. The first the more hardy and sure; the latter the more delicate and luscious. Indeed, few dessert fruits can outmatch a well-ripened, sun-freckled, fully developed and closely compacted bunch of the Diana grape.

The Catawba has its advocates, and it is really a dainty fruit if it have good range of sun, and is not hurried in its ripening; but in delicacy of flavor it must yield to the Diana. The Catawba crop is also exceedingly uncertain in this latitude, by reason of the shortness of the season. A gaunt old vine of this variety, which stands behind the farmhouse, has given me only two crops in the six years past; the frosts have garnered the promise of the others. I have now, however, contrived to conduct its trailing mantle upon a rude trellis, so as to completely embower the roof of the little outlying kitchen; and the fumes and warmth of this latter, from its open skylights, have given to the old vine such a wonderful vigor and precocity, that I have promise of a full burden of well-ripened fruit in advance even of the Isabella. Can the reek of a kitchen be put to better service?

The Isabella escapes ordinary frosts, and is a prodigious bearer; but it has no rare piquancy of flavor; and the same is to be said of its earlier congener, the Hartford-Prolific.

Of all fruits, the grape is the one which, to insure perfection, will least tolerate neglect. I do not speak of those half-wild and flavorless crops, which hang their clusters up and down old elms, in neglected farm-yards,—but of that compact, close array of sunny bunches, where every berry is fully rounded, and every cluster symmetrical. It must have care in the planting, that its fibrous roots may take hold readily upon their new quarters; care in position, which must,—first of all, be sheltered—next, have ample moisture—next, be utterly free from stagnant water, whether above ground or below—and finally, have fair and open exposure to the sun. It must have care in the training, that every spur and cluster may have its share of air and sunshine; care in the winter pruning, to cut away all needless wood; care in the summer pruning, to pinch down its affluence—to drive the juices into the fruit, and to restrain the vital forces from wasting themselves in a riotous life of leaves and tendrils.

But the care required is not engrossing or fatiguing. Any country-liver may bestow it upon the score of vines which will abundantly supply his wants, without feeling the task. Nay, more; this coy guidance of the luxuriant tendrils,—this delicate fettering of its abounding green life,—this opening of the clusters to the gladness of the sunshine, will make a man feel tenderly to the vine, and breed a fellowship that shall make all his restraints, and the plucking away of the waste shoots, seem to be mere offices of friendship.

There is not, anywhere, a country house about

which positions do not abound, where a vine may clamber, and feed upon resources that are worse than lost. The southern or eastern front of an old out-building; a staring, naked wall (on which grapes ripen admirably); a great unseemly boulder, from under which the rootlets will pluck out the elements of the fairest fruit; a back-court, where washings of sinks are wasting; the palings of a poultry-yard—all these are positions, where, with small temptation, the mantling-vine will "creep luxuriant."

I have not alluded to the Delaware, because, thus far, my plants have been poor ones, and my experience unsuccessful. At best, however, the vine is of a more delicate temper than those named, and requires larger care and richer dressing. Under these conditions, I believe the grape to be all, which its friends claim—of a delicate and highly aromatic flavor,—so early as to be secure against frosts, and giving a better promise than any other, of a really good domestic wine.

I am surprised to find in the course of my drives back in the country, how many of our old-time farmers are applying themselves, in a modest and somewhat furtive way, to wine-making. It is true that they bring under contribution a great many foxy swamp varieties, and are not over-careful in regard to ripeness; but faults of acidity they correct by a heavy sugaring, which gives an innocent and bouncing percentage of alcohol.

The practice is not, I fear, entered upon with a purely horticultural love, and I suspect they bring a more lively stomachic fondness to it, than do the pomologists to their science of fruiting. I think the development of this home manufacture has been quickened by Maine-laws, heavy import duties, and by a growing reluctance on the part of the heads of families—to carry a demijohn in the wagon. I also hear the home product commended by the old gentlemen manufacturers, as "warming to the in'ards;" and in large doses, I should think it might be. Their town customers for this beverage are mostly exceedingly serious and sedate people, who have a comical way of calling homemade wines—"pure juice."

And pray, why should not sedate people enjoy the good things of life,—call them by what names they will? I know an exceedingly worthy man who never buys his cider except of a deacon; and then only by the cask; and he buys it very often.

Plums, Apricots, and Peaches.

I AM sorry to give so poor an account, as I needs must, of these stone-fruits. As respects the plum, there is, indeed, an incompatibility of soil upon my farm, to be contended against; but this difficulty is trifling, in comparison with the mischiefs of the arch-enemy, the curculio. The few trees which I found suffering under black-knot in its most aggravated form, I am sorry to say, died under surgical treatment. Others have been planted to supply their places;—planted in the poultry yard—planted in positions where the earth would be hard trampled,—planted in shelter and out of shelter; but although showing fair vigor, and a pretty array of blossoms, no device thus far adopted has succeeded in arresting the spoliations of the curculio. Paving the ground is vain; the forage of poultry is vain; underlying water is vain; and there remain only three resources—to jar off the vermin, gather them and kill them; or second, to deluge the young fruit with a wash that shall nauseate the enemy; or third, to shield the trees or fruit with a gauze covering, that shall forbid attack. They are good devices against any enemy; but extermination is a slow process; if you nauseate the enemy, you are nauseated in turn; and the gauze protection involves a greater sacrifice than the sacrifice of the fruit.

These reasons, though counting against the plum as a market product, do not, of course, forbid its growth as a luxury,—which, like many other luxuries, must be paid for in fourfold its value.

I would by no means undervalue the plum; least

of all, that prince or princess of plums, *Reine-Claude* (Green-Gage), of which, in the sunny towns along the Loire, I have purchased a golden surfeit for a few sous: when I remember those, and their luscious and cheap perfection, crowning the peasants' gardens, I am a little disheartened at thought of the tobacco washes, and whale-oil soap and syringes, with which we must enter into combat with the curculio, for only a most flimsy supply.

The nectarine is subject to the same blight; and the apricot furnishes only a very dismal residuum of a crop. As an espalier, it is not, I think, so subject to the ravages of the curculio as in its unfettered condition; but upon the wall (particularly if one of southern exposure), it is exceedingly liable to injury from the late frosts of Spring. I succeed in saving a few from all enemies every year; but they are so wan—so pinched, as hardly to serve for souvenir of the golden Moor-parks which crown an August dinner at Vefours or the *Trois-Frères.* It is an old fruit; the Persians had it; the Egyptians have gloried in it these centuries past; Columella names it in his garden poem; and Palladius advises that it be grafted upon the almond: * will the nurserymen make trial?

* It occurs in Tit. vii., Novem., where he discourses of the peach. "*Inseritur in se, in amygdalo, in pruno: sed* ARMENIA, *vel* PRÆCOQUA *prunis, duracina amygdalis melius adhærescunt,*" etc.

It will be remembered that in an early chapter I made mention of certain dilapidated peach trees upon the premises, which were even then showing unfailing signs of the 'yellows.' This vegetable dyspepsia has long since carried them off. Indeed, there are but few belts of land throughout New England where a man may hope successful culture of this fruit. The borer is an ugly enemy to begin with; but with close watchfulness, the attacks of this insect may be prevented. His cousin, the curculio, does not greatly affect the downy cheek of a young peach. Yet still, in the absence of more tempting surfaces, he will leave upon it his Turkish signet. Next, comes a curious, foul twisting of the leaves, due—may be—to some minute family of aphides; but this can be mitigated by judicious pruning; after these escapes, and when your mouth is watering in view of a luscious harvest, there appear symptoms of a new disease; the leaves cease to expand; the fruit takes on a premature bloom, and a multitude of little shoots start here and there from the bark, being weakly attempts to struggle against the consuming 'yellows.' And if all these difficulties be fairly escaped or overcome, there remains the damaging fact, that in three winters out of five, in most New England exposures, the extreme cold will utterly destroy the germ of the fruit buds.

Notwithstanding these drawbacks, however, I continue to put out from year to year, a few young trees; not making regular plantations, but dotting them about, in shrubberies, and in unoccupied garden corners, grouping them in the lee of old walls —in the poultry yard,—upon the north side of buildings,—in every variety of position and of soil. In this way I contrive—except the January temperature shows ten below zero—to secure a fair table supply. Even amid the shrubbery of the lawn, where I counted their bloom and foliage a sufficient return, there have been gathered scores of delicious peaches.

I know that it is disorderly, and shocking to all the prejudices of the learned, to plant fruit trees in this hap-hazard way. But I love these offences against system (particularly when system is barren of triumphs). I love to test Nature's own ruling, and give her margin for wide demonstration.

The Poultry.

I KNOW not whether to begin my discourse of poultry with a terrific onslaught upon all feathered creation, or to speak the praises of the matronly fowls, which supply delicate spring chickens to the table, and profusion of eggs. When, on some ill-fated day, a pestilent, pains-taking hen, with her brood of eager chicklings, has found her way into my hot-bed, and has utterly despoiled the most cherished plants; or a marauding drove of young turkeys has cropped all the late cauliflowers, I am madly bent upon extermination of the whole tribe.

But reflection comes—with a nice fresh egg to my breakfast, or a delicate grilled fowl to my dinner—and the feathered people take a new lease of life. They give a sociable, habitable air, moreover, to a country dwelling. The contented, good-humored cluck of the hens, breeds contentment in the on-looker. They are rare philosophers, taking the world as they find it;—now a blade of grass, now a lurking worm; here a stray kernel of grain, and there some tid-bit of a butter-fly; taking their siesta with a wing and a leg stretched out in the sun, and like the rest of us, warning away from their own feeding ground, birds less strong than themselves, with an authoritative dab of their bills. Although amenable to laws of habit, —traversing regular beats for their supply of wild food, and collecting at regular hours for such as the mistress may have to bestow, they are yet rebellious against undue or extraordinary show of authority. It is quite impossible to exercise any safe control over the locality where the hens choose to execute their maternal duties. They insist upon freedom of the will in the matter, as obstreperously, and, I dare say, as logically, as ever any old-school dialectician in his metaphysical homilies.

Nothing could be more charming than the arrangement, matured with the co-operation of an ingenious country carpenter, by which my fowls were to lay in one set of boxes, carefully darkened, and to carry on their incubation in another set of boxes, made cheery (against the long confinement), with sky-light; there were admirable little architectural galleries through which they were to promenade in the intervals of these maternal duties—adroit disposition of courts, and feeding troughs, so that there should be no ill-advised collision,—but it was all in vain. Hens persisted in laying where they should not lay, and in setting, with badly-directed instinct, upon the dreariest of porcelain eggs. The fowls of my Somersetshire neighbor, meantime, at the stone cottage, with nothing more orderly in the way of nests, than a stray lodgment in the haymow, or a castaway basket looped under the rafters of a shed, brought out brood after brood, so full, and fresh, and lusty, as to put my architectural devices to shame.

At certain times, when the condition of the garden or crops allow it, I permit my fowls free forage; and as they stroll off over the lawn and among the shrubberies, it sometimes happens that they come in contact with the more vagabond birds of the larger farm family. The hens take the meeting philosophically, with a well-bred lack of surprise, and are not deterred for a moment from their forage employ; perhaps, (if with a brood) giving an admonitory cluck to their chicks, to keep near them,—even as old ladies with daughters, in a strange place, advise caution, without enjoining positive non-intercourse.

The ducks, on the contrary, in a very low-bred manner, give way to a world of surprises, and gad about each other, dipping their heads, and quacking, and bickering, like old gossips long time apart, who pour interminable scandal in each other's ears. The cocks make an honest, fair fight of it, and one goes home draggled, confining himself thereafter to his own quarters.

The Turkeys meet as fine ladies do, tip-toeing round and round, and eyeing each other with earnest scrutiny, and abundant curvetings of the neck—very stately, dignified, and impudent—stooping to browse perhaps (ladies sniff thus at vinaigrettes), as if no strange fowl were near,—which is merest affectation. They summon their little families into close order, as if fearing contagion, and eyeing each other, wander apart, without a sign of companionship, or a gobble of leave-taking.

I must not forget the groups of Guinea-fowl, who

fraternize charmingly, and threaten to become one family. These birds, unlike all other feathered animals, show no marked difference of appearance between the sexes; so slight is this indeed, that even the naturalists have blundered into errors, and left us in the dark.* Even a fighting propensity does not distinguish the cock, I observe; for the female bird is an arrant termagant, and has undertaken, in my own flock, a fierce battle with a tom-turkey, in which, though worsted, and eventually killed, she showed a fine chivalrous pluck. They are not, however, quarrelsome among themselves; although flocking together in communities, the male birds are strictly faithful to their mates, and manifest none of the sultanic propensities which so deplorably mislead the other domestic fowls.

Notwithstanding their harsh cry, to which the Greeks gave a special descriptive name,† I like the Guinea-fowl; they are excellent layers, enormous devourers of insects—a little over-fond, it is true, of young cauliflowers, and grapes,—yet a stanch, lively, self-possessed bird; and notwithstanding the sneers of Varro,‡ whose taste must have been poor in the matter of poultry,—excellent eating.

* Buffon; *De la Pintard.*

† Καγκάζειν.

‡ Lib. III., De Re Rust. Hæ novissimæ in triclinium ganearium introierunt è culina propter fastidium hominum. Veneunt propter penuriam magno.

The young Guineas, like the young turkeys, are delicate however, and suffer from sudden changes of temperature. Give them what care you will, and all the dietetic luxuries of the books, and on some fine morning, you shall find the half of a brood moping and staggering, and drooping out of life. The young turkeys are even more subject to infantile ailments, and their invalid caprices outmatch all the nostrums of the doctors. Yet some old spectacled lady in the back country, with nothing better than a turned-up barrel in the way of shelter, will by an easy and indescribable 'knack' of treatment, rear such broods as cannot be rivalled by any literal execution of the rules of Boswell and Doyle.

Beyond the age of six weeks, however, danger mostly ceases, and the poults have a good chance—barring the foxes—of coming to the honors of decapitation; and I know few prettier farm sights, than a squadron of pure white turkeys, marching over new mown grass-land, with their skirmishers deployed on either flank, and rioting among the grasshoppers. It is essential that both Guinea-fowl and Turkeys have free and wide range; they are natural wanderers; my hens submit to a curtailment of their liberties with more cheerfulness; but there is after all, no biped of which I have knowledge, that does not glory in freedom. The Black Spanish

9

fowls, Dorkings, and Polish top-knots, (for these make up my variety, and are, I believe, the best), form no exception; and if confinement is necessary, the enclosing palings should be of generous width. A safe rule is—to make the enclosure so large (whatever the number of the flock), that the fowls will not wholly subdue the grass, or forbid its healthful vegetation. If too small for this, it is imperatively necessary for thrift, that they have a run of an hour each day before sunset.

The oldest English writer upon the subject of poultry was a certain Leonard Mascall, who wrote about the year 1581—when Queen Mary was fretting in her long confinement, and Sir Francis Drake was voyaging around the world. He had been farmer to King James, and calls his little black-letter book, "The husbandrye, ordring, and governmente of poultrie." Among his headings are "How to keepe egges long,"—"How to have egges all winter,"—"Of hennes that hatches abroad, as in bushes,"—"Of turquie hennes, profite and also disprofite."

For winter eggs, he advises "to take the croppes of nettles when ready to seed, dry them, and mix them with bran and hemp-seed, and give it to the hens in the morning, and also to give them the seedes of cow-make," (whatever that may be.) I have never ventured trial of his advices; but find full supply in giving hens warm quarters—a closed house, with double walls, and its front entirely of glass; here, with water constantly running, an ample ash box and gravel bed, full feeding,—not forgetting scraps of meat, and occasional vegetable diet—the hens make a summer of the winter, and reward all care. If the weather be very warm, they are allowed a little run in the adjoining barn-yard (their winter home being, in fact, a rustic transmutation of an ancient cow-shed). Any considerable chilliness of the atmosphere, however,—if they are long exposed to it,—checks their laying propensities, and two or three days of housing are needed to restore the due equilibrium.

The Roman writers give us cruel hints in regard to the fattening of fowls, which I have never had the heart to try. They go beyond the rules of the Strasburg poulterers in harshness; and that elegant heathen Columella, has the effrontery to advise that the legs of young doves be broken, in order to cram them the more quickly. Such suggestions belonged, of right, to a period when Roman ladies—Sabina, and Delia, and Octavia—looked down coolly on gladiators, gashing their lives out with bare sabres, and then lolled home in chariots, to dine on thrushes, fatted in the dark. We,—good Christians that we are,—shudder at thought of such barbarism; we pit no bare-backed gladiators against each other, with drawn swords, in our very presence; but we send armies out, of a hundred thousand in blue and gray, and look at their butchery of each other, very coolly, —through the newspapers,—and dine on *pâté de fois gras.* Of course we have improved somewhat in all these ages, since Columella broke pigeons' legs; of course we are civilized; but the Devil is very strong in us still.

Is it Profitable?

WHEN I have shown some curious city visitor all these belongings of the farm—have enlisted his admiration for my crested, golden, Polish fowls,—for my garden, for the fruits;—for the wide stretch of fields, and the herd of cows loitering under the shadow of the scattered apple trees, he turns upon me, in his city way, with the abrupt questioning, "Isn't it confoundedly expensive, though, getting land smoothed out in this style—what with your manures, and levelling, and planting trees?"

And I answer—"N—n—no; no; (somewhat bolder.) There's a certain amount of labor involved, to be sure, and labor has to be paid for, you know. But there are the vegetables, the chickens, the eggs, the milk, and the fruit, which must come out of the shops, unless a man have a home supply."

"To be sure, you're quite right;" and I think he admitted the observation, as many city people incline to, as a new idea. "But," he added, with an awkward inquisitiveness, "Do you ever get any money back?"

My friend was not a reader of the Agricultural Journals, or he could not have failed to notice the pertinacity with which the profitableness of farming is urged and re-urged. Indeed, with all consideration for the calling, I think it is somewhat too persistently pressed. It suggests—rather too strongly the urgence of the recruiting sergeant, in setting forth the profitableness of soldiering. I do not observe that army contractors magnify the gains of their craft very noisily. The hens that lay golden eggs never cackle; at least, I never heard them.

The question of my friend remains however,——

"Do you ever get any money back,—eh?"

What an odious particularity many of these city people have! What a crucial test they bring to the delightful suroundings of a country home! Have they no admiration for such stretch of fields, such herds, and the shrubberies, on whose skirts the flowers are gleaming? Somebody has suggested that the forbidden fruit with which the Devil tempted Eve, and which Eve plucked to the sorrow of her race, was—money. A tree whose fruit carries knowledge of good and evil, is surely not an inapt figure of the capabilities of money—by which all men and women stand tempted to-day. The Paradise tree is not popularly supposed to grow largely on the farms of amateurs.

But the question returns—"Do you get any money back?"

I think it must be fairly admitted that with most amateur farmers, the business (if we reckon it business) is only an elegant luxury; absorbing in a quite illimitable manner, all loose funds at the disposal of the adventurer, and returning—smooth fields, sleek cattle, delicious fruits, and possibly, a few annual premiums. We never get at their 'memoranda.' Mr. Mechi, indeed, of the Tip-tree Hall, gave us an exhibit of his expenses and his sales; but he found it necessary to support the statement with sundry affidavits; people showed wanton distrust; and I think there is an earnest belief among shrewd observers, that the razor straps, soaps, and dressing-cases of Leaden-Hall street (where his original business lies), are, in a large degree, creditors of the Tip-tree Hall farming.

But Mr. Mechi is something more than an amateur; he is an innovator; and has sustained his innovations with a rare business capacity, and that inexorable *system*, which can make even weak ideas exhibit a compacted strength. Amateurs then, cannot take shelter under cover of Mr. Mechi's figures. Farming remains an elegant amusement only, for those who can afford to buy all that they need, and to sell nothing that they raise; and a profitable employment, only (in the majority of instances), for those who can afford to sell all that they raise, and buy nothing that they need.

"Does any money come back, eh?"

The question of my persistent friend must be met. And I do not know how it can be more fairly met, than by giving an abstract of accounts for the first year, third year, and fifth year of occupancy.

Debit and Credit.

LET us count first all extraordinary repairs and necessary implements on taking possession, as part of the farm investment; next, let us set off the interest upon investment, against house rent, and all home consumption. Thus,—if a farm cost $12,000, (and I use illustrative figures only), and if the needed repairs and implements at the start involve an outlay of $3,000 more—we have a total of $15,000, upon which the interest ($900), may be fairly set off against rent, and the poultry, dairy products, fuel, vegetables, etc., consumed upon the place. A shrewd working farmer would say that this implied altogether too large a home consumption, for reasonable profit; but to those who come from the city to the country, with the determination to enjoy its bounties to the full, it will appear very moderate. In any event, it will simplify the comparison I wish to make between the actual working expenses of a farm, and the results of positive sales. But let us come to figures:

FIRST YEAR—EDGEWOOD FARM.

Dr.

To Valuation of live stock,	$1,200 00
" Interest on do.,	72 00
" Purchase of new stock,	300 00
" Labor,	1,200 00
" Hay and grain bought,	150 00
" Seeds, trees, etc.,	150 00
" Manures,	250 00
" Wear and tear of implements,	100 00
" Taxes, insurance, and incidentals,	100 00
	$3,522 00

Cr.

By Valuation stock, close of yr.,	$1,400 00
" Sales do.,	250 00
" do. milk,	600 00
" do. butter,	50 00
" do. vegetables,	60 00
" do. fruits,	10 00
" do. eggs and poultry,	25 00
" do. sundries,	75 00
	$2,470 00
" Balance (loss),	1,052 00
	$3,522 00

First years of any adventure do not offer a very appetizing show—least of all the adventure of restoring a neglected farm.

If this record does not prove entertaining to the reader, I can frankly say that he has my heartiest sympathies. The great enormity lies in the labor account—always the enormity in any reckoning of American farming, as compared with British or Continental. It must be remembered, however, that a large proportion of the sum named, went to the execution of permanent improvements, and that two-thirds of it would have been amply sufficient for the exigencies of the farm-work proper

Let us slip on now to the

THIRD YEAR.

Dr.

To Valuation of stock, ...	$1,500 00
" Interest on do., ...	90 00
" Purchase of new stock . . .	200 00
" Labor bills, ...	1,100 00
" Manures, ...	150 00
" Hay and grain bought, . , .	120 00
" Seeds, trees, etc., ...	50 00
" Wear and tear of implements, .	100 00
" Taxes, insurance, and incidentals, . .	100 00
	$3,410 00
" Balance (gain), ...	615 00
	$4,025 00

CR.

By Valuation stock, close of yr., . . .	$1,600 00
" Sales do., ...	200 00
" do. milk, ...	1,650 00
" do. vegetables, ...	250 00
" do. fruits, ...	125 00
" do. poultry, ...	100 00
" do. sundries, ...	100 00
	$4,025 00

This has a more cheerful look, but is not gorgeous; yet the fields are wearing a trim look, and there is a large percentage of increased productive capacity, which if not put down in figures, has yet a very seductive air for the eye of an imaginative proprietor. Two years later the account stands thus:—

FIFTH YEAR.

DR.

To Valuation of stock, ...	$1,700 00
" Interest on do., ...	102 00
" Purchase of new stock, ...	180 00
" Labor bills, ...	1,000 00
" Manures, ...	100 00
" Grain purchased, ...	130 00
" Seeds, trees, &c., ...	60 00
" Wear and tear of implements, .	100 00
" Insurance, taxes, and incidentals, .	120 00
	$3,492 00
" Balance (gain), ...	988 00
	$4,480 00

Cr.

By Valuation stock, close of yr., . .	$1,700 00
" Sales do.	230 00
" do. milk, ...	1,900 00
" do. vegetables, ...	250 00
" do. fruit, ...	150 00
" do. poultry, ...	130 00
" do. sundries, ...	120 00
	$4,480 00

These figures though written roundly, are (fractions apart) essentially true; I would match them for honesty (though not for largeness), against any official report I have latterly seen—not excepting the "Quicksilver mining," or the Quartermaster General's.

If we analyze these accounts, we shall find the

Average interest upon investment, (say)	$1,000 00
Average working expenses, . .	1,800 00
Total, ...	$2,800 00

On the other side the

Average cash sales are, . .	$2,600 00
House rent and home consumption,	900 00
Total, ...	$3,500 00

Leaving profit of $700, which is equivalent to ten per cent. upon the supposed capital;—all this, under the cheerful hypothesis that personal supervision is a mere amusement, and is not justly chargeable to the farm. If otherwise, and the overlook be rated as Government or corporate officials rate such service, the credit balance becomes ignominiously small.

It is to be considered, however, that the growing productive capacity of the soil, under generous management, may be estimated at no small percentage; and the inevitable increase in value of all lands in the close neighborhood of growing towns, may be counted in the light of another percentage.

All this is not certainly very Ophir-like, nor yet very dreary.

Again, it is to be remarked that the entries for labor, and incidental expenses in the accounts given, are for those expenses only, which contributed directly either to the farm culture, or conditions of culture—not all essential, perhaps, but all contributory. If, however, the Bucolic citizen have a taste to gratify—in architectural dovecots, in hewn walls, in removal of ledges, in graperies, in the planting of long ranges of Osage-Orange (which the winter mice consume), the poor little credit balance of the farm account, is quite lost in the blaze of agricultural splendor.

I do not at all deny the charm of such luxuries. I only say—that they are luxuries; and in the present state of the butter and egg markets, must be paid for as such. And the life that is lived amid such luxuries is not so much a farm life, as it is a life —a long way from town.

Rus hoc vocari debet, an domus longé?

It is the irony of Martial in the concluding line of his Faustine epigram; and with it, I whip my chapter of figures to a close.

Money-Making Farmers.

WHERE are the men then, who have grubbed out of the reluctant eastern soil, their stocking-legs of specie, and their funds at the bank? They are not wholly myths; there are such. Find me a man, who, by aptitude at bargaining (let us not call it jockeyism), can reduce the labor estimates in the foregoing accounts by a third; and who, by a kindred quality, can add to the amount of sales by a third; who can, by dint of early rising and perpetual presence, stretch the ten-hour system into twelve or fourteen; who, by a conquest of all finer appetites, can reduce the home consumption to a third of the figure named in my estimates, and you have a type of one class. A union of tremendous energy and shrewdness; keenly alive to the phases of the market; an ally of all the hucksters; sharp to pounce upon some poor devil of an emigrant, before he has learned the current rate of wages; gifted with a quick scent for all offal, which may be had for the cartage, and which goes to pig food, or the fermentation of compost.

I think I have hinted at a character which those will recognize, who know the neighborhood of large New England towns: a prompt talker—not bashful, —full of life—selectman, perhaps; great in corner groceries, 'forehanded,' indefatigable, trenchant, with an eye always to windward.

If I were to sketch another type of a New England farmer, who is, in a small way, successful,—it would be a sharp-nosed man, thin, wiry, with a blueness about the complexion, that has come from unlimited buffetings of northwesters; one who has been 'moderator' at town meetings, in his day, and upon school committees over and over; one who has sharpened his tongue by occasional talk at 'society' meeting—to say nothing of domestic practice.

I think of him as living in a two-story, white house, with green blinds, (abutting closely upon the road,) and whose front rooms he knows only by half-yearly summations to a minister's tea-drinking, or the severer ordeal of the sewing circle. His hands are stiff and bony; all the callosities of axe and scythe and hoe, have blended into one horny texture, the whole of the epidermis; yet his eye has a keen shrewd flash in it, from the depths of sixty years;

and under the hair of his temple, you may see a remaining bit of bleached skin, which shows that he was—fifty odd years ago—a fair complexioned boy.

He has grown gray upon his straggling farm of one or two hundred acres; yet it is doubtful if the farm will produce more now, than on the day he entered into possession. Some walls have been renewed, and the old ones are tottering. Broken bar-ways have been replaced by new ones; the wood pile has its stock year after year; and every tenth year, when oil is down, the house has its coat of paint—himself being mixer and painter—save under the eaves, for which ladder work, he employs a country journeyman, who takes half pay in pork or grain. When 'help' is low, he clears some outstanding rye field, and commences a new bit of wall —a disunited link, which possibly his heirs may complete. Every year, six, ten, or twelve hogs grow into plethoric proportions; every year they are butchered, under a great excitement of hot water, lard-tryings, unctuous fatty smells—sausage stuffing, and sales to the 'packer' of the town. Every year he tells their weight to his neighbors, between services, at meeting—with his thumb and forefinger in the pocket of his black waistcoat, and the same sly twinkle in his eye.

Every spring he has his 'veals'—four, six, ten,

—as the case may be; and every spring he higgles in much the same way with the town butcher, in regard to age, to price, and to fatness. Every summer I see him in black hat and black dress coat, on his wagon box, with butter firkins behind (the covers closed on linen towels by the mistress at home), driving to the market. And if I trot behind him on his return, I see that his exchange has procured him a two-gallon jug of molasses, a savory bundle of dried codfish, a moisty paper parcel of brown sugar, a tight little bag of timothy seed, and a new hoe, or dung fork. But he never allows his spendings to take up the gross sum of his receipts; always there goes home a modicum, which grows by slow and gradual accretions into notes (secured by mortgage), of some unthrifty neighbor, or an entry upon the columns of his book at the Savings.

There is no amateur of them all, who receives as much into a third, for what he may have to sell; nor any one who spends as little, by two-thirds, for what he may have to buy. It is incredible what such a man will save in the way of barter; and equally incredible how rarely he finds occasion to pay out money at all. Yet he is observant of proprieties; his pew-rent at the meeting house, and tax bills are punctually honored. If I bargain with him, he loves deliberation; he has an opinion, but it only appears after long travail, and comparison of views—in the course of which, he has whittled a stout billet of wood to a very fine point. If I address him in the field, he stops—leans on his hoe—and is willing to lavish upon me the only valuable commodity for which he makes no charge, to wit—his time.

Such a farmer repairs his barn promptly, when the sills are giving way; he does not hesitate at the purchase of a 'likely pair of cattle' at a bargain; he will buy occasional bags of guano, upon proof in his turnip patch, or on his winter rye; but if a sub-soil plow is recommended, he gives a sly twinkle to that gray eye of his, and a complimentary allusion to the old 'Eagle No. 4,' which settles the business.

Such men are in their way—money-makers; but rather by dint of not spending, than by large profits. These back-country gentlemen have their families—educated (thanks to our school system);—boys, lank at the first, in short-armed coats, and with a pinch of the vowel sounds in their speech; but they do not linger around such a homestead; they come to the keeping of hotels, or of woodyards on the Mississippi; many are written down in the dead-books of the war.

Our money-saving farmer has his daughter too, with her Chrysanthemums and striped-grass at the door, and her pink monster of a Hydrangea. She has her Lady's book, and her Ledger, and on such literary food grows apace; but such reading does not instil a healthy admiration for the dairy or butter-making; rosy cheeks and incarmined arms do not belong to the heroines of her dreams. I do not think she ever heard of Kit Marlowe's song:—

"Come live with me, and be my love."

The faint echoes of the town in fashion plates and sensation stories, make a weird, intoxicating music, in listening to which, in weary bewilderment, she has no ear for a brisk bird-song. No wild flowers from the wood, are domesticated at her door. I catch no sight of sun bonnets, or of garden trowels. Out of door life is shunned; and hence, come sallowness, unhealthiness, narrowness—not even the well-developed physique of the town girl, who has the pavements for her marches and countermarches. I hear, indeed, in summer weather, the tinkle of a piano; but it frights away the wrens; and of the two, I must say that I prefer the wrens.

All this unfits for thorough sympathy with the every day life of the father; and when common sympathies do not unite a family, its career breaks at the death of the patron. If there be nothing in the country life which can call out and sustain the pride of all members of the country family, it can never offer tempting career to the young.

From these causes it is, that Dorothy will very likely be a weazen-faced old maid, hopeful of anything but the tender longing of Overbury's "Faire Milke-Maide." Too instructed to admire the sharp roughnesses of her wiry papa; too liberalized, it may be, by her reading, to bear mildly his peevish closeness; not kindling into a love of the beauties of nature, because none will sympathize with that love —dreaming over books that carry her to a land of mirage, and make her still more unfit for the every day duties of life;—not recognizing the heroism of successful struggle with mediocrity and homely duties;—yearning for what is not to be hers, she is the ready victim of illnesses against which she has neither the vigor nor the wish to struggle.

"So, Dorothy is gone! Squire," says the country parson; "Let us pray to God for his blessing."

The darkened parlors are opened now; the farmer's daughter is a bride, and death is the groom.

The gilt-backed books are dusted; the cobwebs swept away; the black dress-suit rebrushed; the twinkle of the eye is temporarily banished; the neighbors are gathered; the warning spoken; the procession moves; and the grave closes it all.

The Artemisias bloom on, and the purple tufts of Hydrangea;—poor Dorothy's flowers!

It is a little picture from the life of certain money-making farmers, who pinch—to save. There is a jingling resonance of money at the end, but it is not tempting; it has come upon a barren life, without glow or reach—a life whose parlors have been always closed.

Does Farming Pay?

AND now let us *prèciser* the whole matter, and get rid, if we can, of that interminable question—does Farming pay?

Will shop-keeping pay? Will tailoring or Doctoring pay? Will life pay? How do these questions sound? And yet they are as reasonable as the one we come to consider. Tell me of the capacity of the Doctor—of the tailor; tell me of his location, and of his aptitude for the business, and I can answer. Tell me of what material you propose to make a farmer, tell me of his habits, and of the condition of his soil and markets, and I can tell you if he will find a profit or none; and this, without regard to Liebig, Short-horns, or the mineral theory.

Successful farming, it must be understood, is not that which secures a large monied result this year, and the next year, and the year after; but it is that which ensures to the land a constantly accumulating fertility, in connection with remunerative results. The theory of the agricultural doctors, that every year, as much of the nutritive elements of land should be restored, as the annual cropping removes, may be good ruling for virgin soil, or for the Lothians, or Belgian gardens; but for neglected or poor soil, a larger restoration is needed;—if not by manures, then by tillage or drainage. Exact equipoise is difficult, and implies no advance. It is neither easy nor desirable to be forever balancing oneself upon a tight rope. If progressive farming will not pay, it is quite certain that no other farming will.

I know there are many quiet old gentlemen among the hills, who have a sleepy way of putting in their corn patch year after year, and a sleepy way of clearing out their meagre pittance of drenched manure, and a sleepy way of never spending, who drop off some day, leaving money in their purse; but such success does not tempt the young; it gives no promise of a career. "Pork and cabbage for dinner, and the land left lean,"—might be written on their gravestones.

The faculty of not-spending, is cultivated by many farmers, a great deal more faithfully than their lands; but the faculty of right-spending (*facultas impendendi*),* is at the bottom of all signal

* The language of Columella, which is as keen and as much to the point now as in the time of Tiberius:—"*Qui studium agricolationi dederit, sciat hæc sibi advocanda: prudentiam rei, facultatem impendendi, voluntatem agendi.*"

success in agriculture, as in other business pursuits. This kind of enterprise is what farmers specially lack; and the lack is due to the secure tenure by which they hold their property. The shopkeeper who turns his capital three or four times in a year, and who knows that an old stock of goods will involve heavy losses, is stimulated to constant activity and watchfulness. The farmer, on the other hand, inheriting his little patch of land, and feeling reasonably sure of his corn and bacon, and none of that incentive which attends risk, yields himself to a stolid indifference, that overlays all his faculties. Yet some of the Agricultural papers tell us with pride, that bankruptcies among farmers are rare. Pray why should they not be rare? The man who never mounts a ladder, will most surely never have a fall from one. Dash, enterprise, spirit, wakefulness, have their hazards, and always will; but if a man sleep, the worst that can befal him is only a bad dream. This lethargy on the part of so many who are content with their pork dinners and small spendings, is very harmful to the Agricultural interests of the country. Young America abhors sleepiness, and does not gravitate, of choice, toward a pursuit which seems to encourage it. The conclusion and the conviction have been, with earnest young men, that a profession which did not stimulate to greater activity and larger triumphs, and a more Christian amplitude of life, could not be worth the following. Nothing about it or in it seemed to have affinity with the great springs of human progress otherwheres; a lumpish, serf life, it seemed—bound to the glebe, and cropping its nourishment thence, like kine.

Again, the extravagance of those who have undertaken farming as a mere amusement, has greatly damaged its character as a pursuit worthy the enlistment of earnest workers. Our friend, Mr. Tallweed, who, with his Wall-street honors fresh upon him, comes to the country to grow tomatoes at a cost of five dollars the dozen, and who puts a sack of superphosphate to a garden row of sweet corn, may make monstrosities for the exhibition tables, but he is not inviting emulation; he is simply committing an Agricultural debauch. And an Agricultural debauch pays no better than any other.

But between these extremes, there is room for a sober business faculty, and for an array of good sense. With these two united, success may be counted on; not brilliant perhaps, for in farming there are no opportunities for sudden or explosive success. The farmer digs into no gold lead. He springs no trap, like the lawyer or tradesmen. His successes, when most decided, are orderly, normal, and cumulative. He must needs bring a cool temper, and the capacity—to wait. If he plant a thousand guineas—however judiciously,—they will not sprout to-morrow. There have been, I know, Multicaulis fevers, and Peabody seedlings; but these are exceptional; and the prizes which come through subornation of the Patent Office, are rare, and dearly paid for.

Again, it must be remembered, that all success depends more on the style of the man, than on the style of his business. For one who is thoroughly in earnest, farming offers a fair field for effort. But the man who is only half in earnest, who thinks that costly barns, and imported stock, and jaunty fencing, and a nicely-rolled lawn are the great objects of attainment, may accomplish pretty results; but they will be small ones.

So the dilettante farmer, who has a smattering of science, whose head is filled with nostrums, who thinks his salts will do it all; who doses a crop—now to feebleness, and now to an unnatural exuberance; who dawdles over his fermentations, while the neighbor's oxen are breaking into his rye field; who has no managing capacity—no breadth of vision,—who sends two men to accomplish the work of one—let such give up all hope of making farming a lucrative pursuit. If, however, a man be thoroughly in earnest, if he have the sagacity to see all over his farm—to systematize his labor, to carry out his plans punctually and thoroughly; if he is not above economies, nor heedless of the teachings of science, nor unobservant of progress otherwheres—let him *work*,—for he will have his reward.

But even such an one may, very likely, never come to his " four in hand," except they be colts of his own raising; or to private concerts in his grounds —except what the birds make.

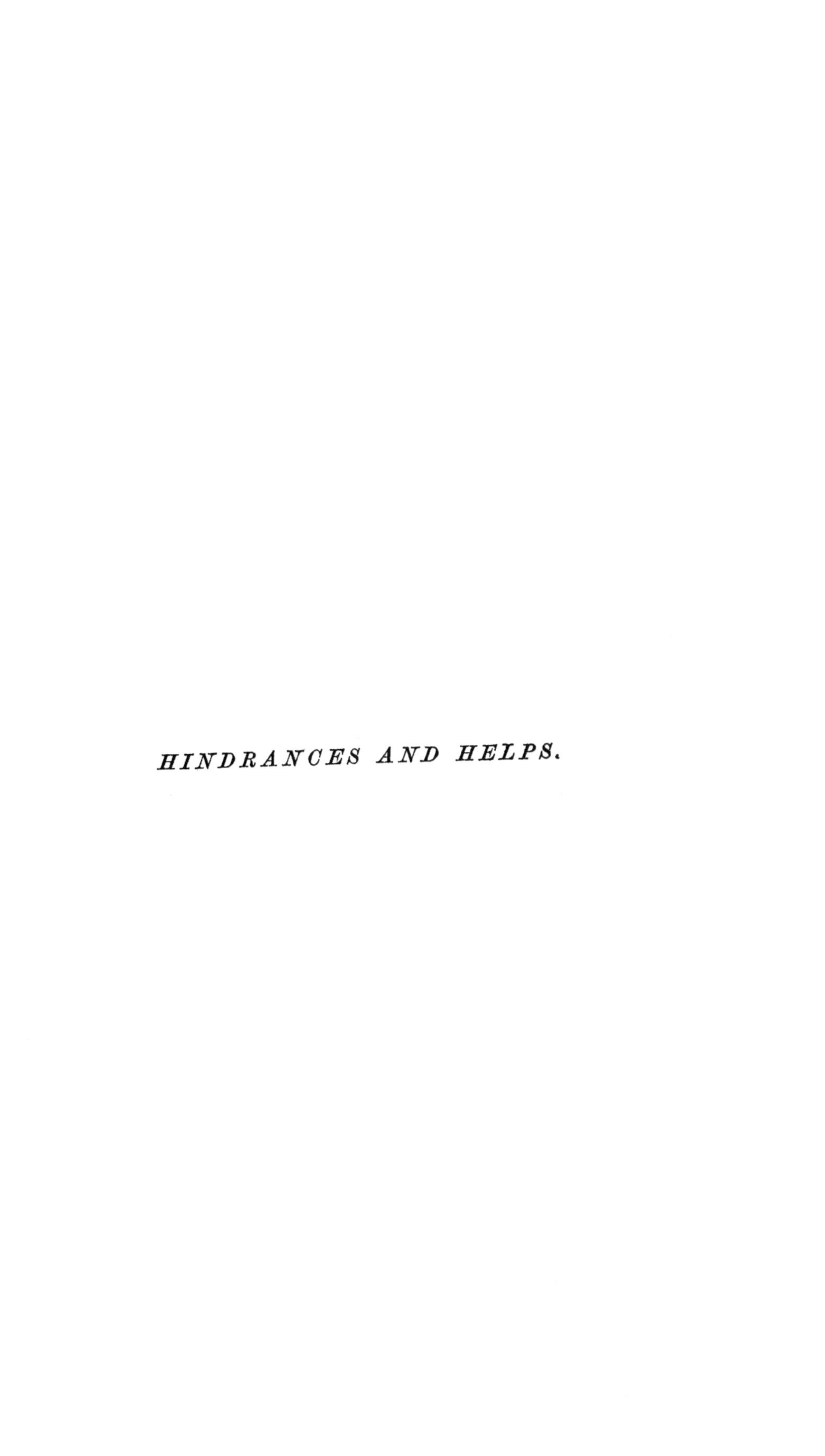

HINDRANCES AND HELPS.

IV.

HINDRANCES AND HELPS.

The Argument.

IT will be perceived by the reader who has been kind enough to follow me thus far, that this book neither professes to be wholly practical, nor yet wholly fanciful. It is—if I may use a professional expression—the fruit from a graft of the fanciful, set upon the practical; and this is a style of grafting which is of more general adoption in the world than we are apt to imagine. Commercial life is not wholly free from this easy union,—nor yet the clerical. All speculative forays, whether in the southern seas or on the sea of metaphysics, are to be credited to the graft Fancy; and all routine, whether of ledger or of liturgy, go to the stock-account of the Practical. Nor is the last necessarily always profit, and the other always loss. There are, I am sure, a great many Practical failures in the world, and the number of Fanciful successes is unbounded.

I have endeavored more especially to meet and to guide, so far as I may, the mental drift of those who think of rural life, either present or prospective,—not as a mere money-making career (like a dip into mining)—nor yet as the idle gratification of a caprice. No sensible man who establishes himself in a country home, desires that the acres about him should prove wholly unremunerative, and simple conduits of his money; nor yet does he wish to drive such a sharp bargain with his land as will cause his home to be shorn of all the luxuries, and the legitimate charms of a country life. It is needless to say that I hope for sensible readers, and direct my observations accordingly. With this intent I propose, in this last division of my book, to review all the helps and hindrances to the success and the rational enjoyment of a farm-life. I shall not reason the matter so closely as I might do, if I were addressing the attendants upon a County-Fair, but shall scatter my hints and experiences through a somewhat ample margin of illustrative text, from which the practical man may excerpt his little nuggets of information or suggestion,—as the case may be; and the reader who is pastorally inclined, may find frequent dashes of country perfume, that shall deftly cover the ammoniacal scents.

Agricultural Chemistry.

WHEN a man buys clean copies of Liebig and of Boussingault, and walks into possession of his land with the books under his arm, and an assured conviction that with their aid, he is about to supplant altogether the old practice, and commit havoc with old theories, and raise stupendous crops, and drive all his old-fashioned neighbors to the wall,—he is laboring under a mistake. His calves will very likely take the 'scours;' the cut-worms will slice off his phosphated corn; the Irish maid will pound his cream into a frothy chowder;—in which events he will probably lose his temper; or, if a cool man, will retire under a tree, and read a fresh chapter out of Liebig.

There are a great many contingencies about farming, which chemistry does not cover, and probably never will. People talk of agricultural chemistry as if it were a special chemistry for the farmer's advantage. The truth is (and it was well set forth, I remember, in a lecture of Professor Johnson's), there is no such thing as agricultural chemistry; and the term is not only a misnomer, but misleads egregiously. There is no more a chemistry of agriculture than there is a chemistry of horse-flesh, or a conchology of egg-shells. Chemistry concerns all organic and inorganic matters; and, if you have any of these about your barn-yards, it concerns them; it tells you —if your observation and experience can't determine —what they are. Of course it may be an aid to agriculture; and so are wet-weather, and a good hoe, and grub, and common-sense, and industry. It may explain things you would not otherwise understand; it may correct errors of treatment; it may protect you from harpies who vend patented manures—not because it is agricultural chemistry; but, I should say rather, looking to a good deal of farm practice—because it is not agricultural, and because it deals in certainties, and not plausibilities. There is such a thing as religion, and it helps, sometimes, to purify Democrats and sometimes Republicans; but who thinks of talking—unless his head is turned—about democratic religion, or republican christianity?

The error of the thing works ill, as all errors do in the end. It indoctrinates weak cultivators with the belief that the truths they find set down in agricultural chemistries, are agricultural truths, as well as chemical truths; and thereupon, they mount a promising one as a hobby, and go riding to the wall. Chemistry is an exact science, and Agriculture is an experimental art, and always will be, until rains stop, and bread grows full-baked. A chemical truth is a truth for all the world and the ages to come;

and if you can use it in the making of shoe-blacking, or to dye your whiskers,—do so; but don't for that reason call it Whisker-chemistry.

It is a chemical truth that an alkali will neutralize an acid if you furnish enough of it; and if, with that truth festering in your brain, you can contrive to neutralize your entire fund of oxalic-acid, so that no sorrel shall thenceforth grow,—pray do so. But I do not think you can; and first, because the soil—to which quarter you would very naturally direct your alkaline attack—may be utterly free of any oxalic acid whatever; its presence in the plant, is no evidence of its presence in the soil. Pears have a modicum of pectic acid at a certain stage of their ripeness, but I suspect it would puzzle a sharp chemist to detect any in the soil of a pear-orchard. And even if the acid were a mineral acid, and were neutralized—it must be remembered—that to neutralize, is only to establish change of condition, and not to destroy;—how know you that the little fibrous rootlets will not presently be laying their fine mouths to the neutral base, and by a subtle alchemy of their own, work out such restoration as shall mock at your efforts—in all their rampant green, and their red tassels of bloom?

The presence of any particular substance in a crop, does not *ipso facto*, warrant the application of the same substance to the soil as the condition of increased vigor. The man who,—having retired to the shade for a fresh chapter of Liebig,—finds that cellulose enters largely into the structure of his plants, and thereupon gives his crops a dressing of clean, pine saw-dust, would very likely have his labor for his pains. That wonderful vital laboratory of the plant, has its own way of effecting combinations; and stealing, as it does, the elements of its needed cellulose, in every laughing toss of its leaves—it scorns your offering.

It is a chemical truth that the starch in potatoes or wheat, is the same thing with the woody fibre of a tree; but it is not an agricultural fact—differs as widely from it in short, as a stiffened shirt-collar from the main-mast of a three-decker ship. A farmer comes to the chemist with some dust or bolus from a far-away place, and asks what is in it; he can tell upon examination, and if, after such examination he finds it to possess a large percentage of soluble phosphoric acid, he will advise its use as a manure, and can promise that it will contribute largely to the vigor of a wheat crop; all this—not simply because phosphoric-acid is a constituent part of the grain, but because he knows that other dressings containing a like element, have invariably so contributed; the fact being established by repeated farm-trials. But it is not a result determinable, so far as a field-crop is concerned, by simple chemical investigation; nor could it be so determinable, unless you could establish the crop and feed it, under those conditions of alienation from all other influences, by which or under which alone, the chemist is enabled to establish the severity of his conclusions.

The power of the chemist to decompose, to unravel, to tear in pieces, and to name and classify every separate part, is something wonderful; but his power to combine is less miraculous. Give him all the carbonic-acid in the world and he cannot make us a diamond, or a lump of charcoal. And when, with the natural combination is associated a vital principle, (as in plants), controlling, amplifying, decomposing at its will, his power shrinks into still smaller dimensions. Faithful and long-continued observation of the mysterious processes of nature, will alone justify a theory of plant-nutrition. A large part of this observation is supplied by the history of farm-experiences, and another part is supplied by the earnest investigations of special scientific inquirers. Where the two tally and sustain each other,—one may be sure of standing upon safe ground. But where they are antagonistic, one has need to weigh conflicting evidence well, not presuming hastily that either practical experience, or a special science has, as yet, a monopoly of all the truths which lie at the base of the "mystery of husbandry." For these reasons it is, that I say,—let no man rashly hope to revolutionize farming, upon the strength of clean copies of Liebig and Boussingault.

A Gypseous Illustration.

THE farming community has a great respect for men of science; it never thinks of distrusting any of their dicta, so long as they are conveyed in scientific and only half-intelligible language. The working-farmer is altogether too busy and shrewd a man to controvert a statement of which he has only vague and muddy comprehension. His dignity is saved, by bowing acquiescence, and passing it unchallenged. Thus,—if the Professor, talking in the interests of agriculture, says: "Gypsum is very serviceable in fixing the ammonia which is brought down from the atmosphere by showers," the common-sense farm-listener is disposed to admit so airy a truth. But if the Professor, meeting him over the fence, says: "Plaster is an excellent manure," the common-sense man retorts:

"Waal—d'n'know; depends a leetle upon the sile, in my opinion."

But as the scientific man confines himself mostly to the language of the desk, and meets with an admiring assent, he is apt, I think, to generalize somewhat too loosely and rashly in his theories of applied science. Naturally enough, confident in the results of his own investigation, he entertains a certain contempt for a merely empirical art; he undervalues the experience and practices of its patrons, and proposes to lay down a law for them, which, having scientific truth for its basis, may work unvarying results. I do not know how I can better illustrate this, than by noticing some of the various theories which have obtained, in respect to the fertilizing action of gypsum.

A farmer, for instance, finds himself within easy reach of a large supply of this salt, and being chemically inclined, he sets himself to the task of reading what has been written on the subject,—in the hope, possibly, of astounding the neighbors, and glutting the corn market.

At the outset I may remark, that farm-experience has as yet found no law by which to govern the application of gypsum; on one field it succeeds; in another, to all appearance precisely the same, it fails; at one time it would seem as if its efficacy depended on showers following closely upon its application; in other seasons, showers lose their effect. In one locality, a few bushels to the acre work strange improvement, and in another, fifty bushels work no change whatever. Now—it is a hill pasture that delights in it, and again—it is an alluvial meadow. Hence it offers peculiarly one of those cases, where an observant and earnest farmer would be desirous of calling in the aid of scientific opinion.

And what will he find?

Sir Humphry Davy, that devout old gentleman, who was as good an angler as he was chemist, exploded the idea prevalent in his day—that gypsum was beneficial by promoting putrefaction of manurial substances—and expressed the opinion that it was absorbed by the plants bodily; at least by those plants whose ash showed large percentage of sulphate of lime. Sir Humphry was honest; the theory was not too absurd; the farmers were doubtless glad to get a handle to their talk about plaster; and so for a dozen years or more, the lucerne and clover went on absorbing the gypsum. At last some inquisitive party ascertained, by careful experiment, that a field of clover not treated with gypsum, contained as large a percentage of sulphate of lime in its ash, as another field which had been treated to the salt. The inference was plain, that the superior vigor of the last was not attributable to simple absorption of the sulphate, and the theory of Davy quietly lapsed.

Chaptal, the French chemist, speaks of gypsum in a loose way as a stimulator; but in what particular direction its stimulating qualities are supposed to work, he does not inform us.

About the year 1840, I think, Dr. Dana, of Lowell, published a bouncing little book called a Muck Manual, in which he affirmed very stoutly that gypsum was quietly decomposed by the roots of the plants, when its sulphuric acid flew off at the silicates, and worried them into soluble shape; and its lime, on the other side, flew off at the *geine*, pounding that into a good relish; in short, he made out so charming a little theory,—so vivacious in its action,—so appetizing to turnips, and so authoritatively stated, that we farmers must needs accept it at a glance, and take off our hats, with—"That's it,"—"I thought so," —"The very thing."

But straight upon this, like a thunder-clap, comes Liebig,* who declares, in *his* authoritative way, that the value of gypsum "is due to its faculty of fixing the small quantity of carbonate of ammonia, brought down by the rain and the dew;" at this, we farmers put on our hats again, and waited for the rain.

Some two or three years after, M. Boussingault, who had gone through the South-American wars under Bolivar, and studied agriculture at Quito, as well as on his own country-estate of Bechelbron,

* His first book appeared in America, if I am not mistaken, in 1841.

entertains us with the report,—in *his* mildly authoritative way, and sustained by great weight of evidence, —that Dr. Liebig is utterly wrong in his theory, and that the value of gypsum is due entirely to the lime which it introduces into the soil;—the sulphuric acid, which played such a lively game under the pen of Dr. Dana—counting for nothing.

By the time this stage of the inquiry is reached, the investigating young farmer, with whom I entered upon this illustration, might be safely supposed to be slightly muddled; and yet, with a comparatively clear recollection of the last-presented theory in his mind, he might farther be supposed to consider the propriety of buying lime at eight cents a bushel, rather than gypsum at sixty cents.

But he has hardly formed this decision, and seen his lime dumped upon his clover-field, when he receives a copy of Dr. Liebig's final work upon the Natural Laws of Husbandry. Turning with nervous haste to the Doctor's discussion of the sulphate of lime, he finds these startling statements: "It may be safely assumed that in cases where gypsum is found to be favorable to the growth of clover, the cause must *not* be sought for in the lime; and since arable soil has the property of absorbing ammonia from the air and rain water, and *fixing it in a higher degree than salts of lime*, there is only the sulphuric acid left to look to for an explanation of the favorable action of gypsum."

And in this muddle I leave our young farmer, contemplating, in an abstracted manner, his lime heap, and reflecting upon the wonders of nature.

Yet it is not altogether a muddle. Science has failed in substantiating a theory of action—only where all farm experience is equally at fault; when the two march together, they pluck up triumphs by the roots. The particular action of gypsum, with a safe rule for its application, remains one of the mysteries of the craft; and there are a great many others. Science is not discredited, however, by the antagonism of such men as Liebig and Boussingault. Stout men will stagger, when they explore the way for us into the dark. The dignity of science will suffer more from the pestilent iteration of smatterers who presume to solve all the riddles of nature in their own little retorts. And the danger is all the greater from the fact that uninstructed farmers render an instinctive respect and confidence to a man who professes familiarity with science. It is never imagined by them, that one who would write $C_8H_4O_8+2HO$ for malic acid,—would tell an untruth or take airs upon himself. Yet I think it may be safely conceded that a rash man, or a mischievous man may cover falsehood under such formulæ, as easily

—as if he edited a morning paper. And I really do not know how I could put the matter more strongly.

With respect to gypsum,—and in close of this special topic,—I may say that I have found it sometimes of service upon young clover, and sometimes of no service at all. Upon old pasture land, it has, with me, uniformly counted for nothing; and again, I have never failed to find an appreciable increase of the crop of potatoes, where I have sown gypsum in the trenches at planting. It is certain that we have no right to condemn the salt, simply because we cannot detect the precise mode of its operation. That mode, I am inclined to believe very complex, and that no uniform law will ever meet the requirements of the case; nor have I a doubt but that in process of time, and under the tests of a future and finer chemistry, and of a fuller experience, every one of the dilute theories named, will throw down its little flocculent precipitate of truth.

Science and Practice.

I REMEMBER once, in company with a crowd of interested auditors, listening to a justly distinguished pomologist, who, in the course of his peroration in praise of scientific study, suggested the great advantage of analyzing all the different pears, and the different soils under culture, so that they might be minutely adjusted each to each. Of course the worthy old gentleman never did such a thing; and (being a shrewd man) never means to. Yet it seemed not a very bad thing—to say. The lesser pomologists all wagged their heads approvingly, but without any serious thought of following the advice; the embryo chemists fairly gushed over in approval; and the only doubt expressed, was in the faces of certain earnest, honest, old farmers,—who had already paid their twenty-five dollars for a soil analysis, to the eminent Professor Mapes,—and of one or two scientific adepts, who, I thought, gave a twirl to their tongues in the left cheek,—rather evasively. In general, I find that the most modest opinions in regard to the agricultural aids of applied science, come from the men of most distinguished scientific attainment; and the exaggerated promises and suggestions flow from those who are slightly indoctrinated, and who make up by uproar of words, and aggregation of pretentious claims, for the quiet confidence and far-sighted moderation of real science. Even so we find a General in command—looking from end to end of the field—modest in his promises, doubtful by reason of his knowledge; while some blatant Colonel, puffy with regimental valor, and knowing the positions only by the confused roar of artillery, will pompously threaten to bag every man of the enemy!

But aside from the exaggeration alluded to,—and of which I should reckon so minute a soil-analysis as to determine what ground would most favor the development of *pectose* in a baking pear, and of *pectic acid* in a Bartlett, a fair sample,—there are other hindrances to the effective and profitable co-laboration of scientific men with the practical farmer. The latter has a wall about him of self-confidence, ignorance of technicals, great common-sense, and awkward prejudices, which the scientific man, with his precision, his fineness of observation, his remote analogies, and his impatience of guess-work, is not accustomed or fitted to undermine. He may breach indeed successfully all the old methods, but if the old methodist does not detect, or recognize the breach, what boots it? Science must stoop to the work, and show him a corn crop that is larger and *grown more cheaply* than his own; this is sending a shot home.

Let me illustrate, by a little talk, which I think will have the twang of realism about it.

A shrewd chemist, devoting himself to the missionary work of building up farming by the aid of his science, pays a parochial visit to one of the backsliders whom he counts most needful of reformation. The backslider,—I will call him Nathan,—is breaking up a field, and is applying the manure in an unfermented and unctuous state ;—the very act of sinning, according to the particular theory of our chemist, perhaps, who urges that manures should be applied only after thorough fermentation.

He approaches our ploughing farmer with a "Good morning."

"Mornin'," returns Nathan (who never wastes words in compliment).

"I see you use your manure unfermented."

"Waal, I d'n'know—guess it's about right; smells pooty good, doan't it ?"

"Yes, but don't you lose something in the smell ?"

"Waal, d'n'know ;—kinder hard to bottle much of a smell, ain't it ?"

"But why don't you compost it; pack up your long manure with turf and muck, so that they will absorb the ammonia ?"

"The what ?—(Gee, Bright !)"

"Ammonia; precisely what makes the guano act so quickly."

"Ammony, is it ? Waal,—guanner has a pooty good smell tew ; my opinion is, that manure ought to have a pooty strong smell, or 'taint good for nuthin'."

Scientific gentleman a little on the hip ; but revives under the pungency of the manure.

"But if you were to incorporate your long manure

with turf and other material, you would make the turf good manure, and put all in a better state for plant food."

"Waal—(considering)—I've made compo's afore now ;—dooz pooty well for garden sass and sich like, but it seems to me kinder like puttin' water to half a glass o' sperit; it makes a drink a plaguey sight stronger'n water, no doubt o' that; but after all 's said and dun,—'taint so strong as the rum. (Haw, Buck; why don't ye haw!)"

Scientific gentleman wipes his spectacles, but follows after the plough.

"Do you think, neighbor, you're ploughing this sod as deeply as it should be?"

"Waal—(Gee, Bright!)—it's as folks think; I doan't like myself to turn up much o' the yaller; it's a kind o' cold sile."

"Yes, but if you exposed it to the air and light, wouldn't it change character, and so add to the depth of your land?"

"Doant know but it might; but I ha'n't much opinion o' yaller dirt, nohow; I kinder like to put my corn and potatoes into a good black sile, if I can get it."

"But color is a mere accidental circumstance, and has no relation to the quality of the soil."

("Gee, Bright! gee!")

"There are a great many mineral elements of food lying below, which plants seek after; don't you find your clover roots running down into the yellow soil?"

"Waal, clover's a kind of a tap-rooted thing,—nateral for it to run down; but if it runs down arter the yaller, what's the use o' bringin' on it up?"

The scientific gentleman sees his chance for a dig.

"But if you can make the progress of the roots easier by loosening the sub-soil, or incorporating a portion of it with the upper soil, you increase the facilities for growth, and enlarge your crops."

"Waal, that's kinder rash'nal; and ef I could find a man that would undertake to do a little of the stirrin' of the yaller, without bringin' much on't up, and bord himself, I'd furnish half the team and let him go ahead."

"But wouldn't the increased product pay for all the additional labor?"

"Doant b'lieve it would, nohow, between you and I. You see, you gentlemen with your pockets full o' money (scientific gentleman coughs—slightly), talk about diggin' here and diggin' there, and turnin' up the yaller, and making compo's, but all that takes a thunderin' sight o' work. (Gee, Bright!—g'lang, Buck!)"

The scientific gentleman wipes his spectacles, and tries a new entering wedge.

"How do you feed your cattle, neighbor?"

"Waal, good English hay; now and then a bite o' oats, 'cordin' as the work is."

"But do you make no beeves?"

"Heh?"

"Do you fatten no cattle?"

"Yaas, long in the fall o' the year I put up four or five head, about the time turnips are comin' in."

"And have you ever paid any attention to their food with reference to its fat-producing qualities, or its albuminoids?"

"(Gee, Bright!)—bumy—what?"

"Albuminoids—name given to flesh producers, in distinction from oily food."

"Oh,—never used 'em. Much of a feed? (G'lang, Buck!)"

"They are constituent parts of a good many varieties of food; but they go only to make muscle; it isn't desirable you know to lay on too much fatty matter."

"Heh?—keep off the fat do they? (Gee, Bright!) Dum poor feed, then, in my opinion."

By this time the end of the furrow is reached, and the scientific gentleman walks pensively toward the fence, while Nathan's dog that has been sleeping under a tree, wakes up, and sniffs sharply at the bottom of the stranger's pantaloons.

I have written thus much, in this vein, to show the defensible position of many of the old style farmers, crusted over with their prejudices—many of them well based, it must be admitted—and armed with an inextinguishable shrewdness. The only way to prick through the rind is to show them a big crop grown at small cost, and an orderly and profitable method, gradually out-ranking their slatternly husbandry. Nor can I omit to say in this connection, that the free interchange of questions and answers, and unstarched companionship, which belonged to the New Haven Agricultural Convention of 1860, are among the best means of breaking down the walls of demarcation, and establishing chemical affinities between Science and Practice.

Lack of Precision.

THE manufacturer, in ordinary times, can tell us with a good deal of certainty how much work he can turn out in any given month, and what his profits will be. The farmer, whose crops are dependent in a greater or less degree upon contingencies of wet, or rain, or cold, over which he has no control, is less positive; and as a consequence, I think, he grows into an exceedingly loose habit of thought in all that regards his affairs. Notwithstanding his punctiliousness in moneyed details, and his sharpness at a bargain, he has a more vague idea of his real whereabouts in the world of profit and loss, than any man of equal capital that you can find. If he has a little pile in stocking-legs or in Savings that grows,—it is profit; if he has a little debt at the grocer's or the bank that grows,—it is loss.

There is not one in fifty who can tell with anything approaching to accuracy, how much his grain or roots cost him the bushel; not one in fifty who can show anything like a passable balance sheet of a year's transactions. He may put down all the money he receives in stumpy figures, and all the money he pays out in other stumpy figures, and set his oldest boy to the Christmas reckoning. But his rent, his personal labor, the wear and tear, the waste, the consumption, the unmarketed growth, assume only a hazy indeterminate outline, within which the sum of the stumpy figures is lost. Whether he is raising corn at a price larger than the market one, or selling potatoes for a third less than they cost him, is an inquiry he never submits to the fatigue and precision of accurate investigation. He thinks matters are *about* so and so; his oxen are worth *about* so much; his oats will turn *about* thirty bushels to the acre.

Nay, he carries this looseness of language into matters of positive knowledge; the straightest stick of timber in the world is only *about* straight, and the politicians are *about* as dishonest as they well can be.

Suppose we try him upon his corn crop; we submit that it looks a little yellow.

"Waal—yes, kinder yeller; t'ain't fairly caught hold o' the dung yit" (pegging away with his hoe).

"Do you think there's any profit in growing corn, hereabout?"

"Waal—don't know as there is much; kinder like to make a little pork, and have a little about for the hens."

"But why not buy your corn and raise something else, provided you can buy it, as you often can, for sixty or seventy cents the bushel."

"Waal—kinder like to have a little 'heater' piece; the boys, you see, hoe it out in odd spells; don't pay out much for help."

"But the boys could earn their seventy-five cents a day, couldn't they?"

"Waal—suppose they might—about; but kinder like to have 'em about home."

"Have you ever tried carrots?"

"Waal—no; kinder back-achin' work to weed carrits."

And not only does this apathetic indifference to

the relative profits of different crops prevail, but there is no proper business estimate of home labor.

We often see it affirmed, admiringly, that such or such a man has built an enormous quantity of wall —so many feet high and broad—or dug out so many rocks, and mostly with his own hands, or in spare time with his own 'help'; in short, it is intimated that all is done at little expense. Now this is very absurd; great work involves great labor; and great labor has its price. You may do it in the night, and call it no labor; you may do it yourself, and call it no expense; but there is, nevertheless, a great deal of positive expenditure of both muscle and time which, if not given to this work, might have been given to another. It may count much for your industry, but not one whit for your farming, until we learn if the labor has been judiciously expended—has paid, in short. And to determine this, we must estimate the labor at its market value—whether done in the night, or on holidays.

If I see a house painted all over in diamonds of every hue, and express distaste for the wanton waste of labor, it is no answer to me to say—that the man did it in odd hours. What will not pay for doing in even hours, will never pay for doing in odd hours. It is no excuse for waste of time and muscle, to waste them in the dark. Every spade or hammer-stroke upon the farm—no matter whether done by the master or the master's son, or master's wife—no matter whether done after hours or before hours—must be estimated at the sum such labor would command in the market.

The fallacy is only another indication of that woful lack of precision of which I have been speaking, and which, I am sorry to say, infects more or less the current Agricultural literature. A well-meaning man gives some account of an experiment that he has undertaken, and is so loose in statement of details, so inexplicit, so neglectful to make known previous conditions of soil, or conditions of cost, that he might as well have burst a few soap-bubbles in the face of the public.

Even in reports of State societies, the estimate of labor and other expenses on premium-crops is so various, so conflicting, often so patently and egregiously wrong, that it is quite impossible to arrive even at a safe average. I find among these reports, the calculation of some short-figured farmer, who has competed for a premium upon his carrots, and who has the effrontery to put down the cost of cultivating and harvesting an acre—at twenty dollars! Yet he won his premium, and the estimate stands recorded. The committee who audited and accepted such a report—if donkeys were on exhibition—should have been put round the track.

Knowing too Much.

I SOMETIMES see in the papers, advertisements of gardeners, who can be seen at Thorburn's, in John street, on stated mornings, when they hold their levee, who insist upon 'entire control.' A modest man, going among them, and entreating the services of one at forty dollars a month, and 'boord,' feels very much as if he were hiring himself to him in some subordinate capacity,—with the privilege of occasionally sniffing the perfume through the open doors of the green-house. There may be those country-lovers who enjoy this state of dependence upon the superior authority of a gardener; but I do not care to be counted among them. I have too large an acquaintance among the sufferers. M——, an amiable gentleman, and a friend of mine, and an extreme lover of flowers, dared no more to pick a rose without permission of 'Wallace,' than he dares to be caught reading an unpopular journal. 'Wallace' is instructed; but in the assertion of his authority, —impudent. And when at last my friend summoned resolution to dismiss him, there came a dray to the back-entrance, which was presently loaded down with the private cuttings and perquisites of the accomplished gardener.

When a gardener knows so much as to refuse any suggestions, and to disallow any right on the part of the proprietor to stamp his place with his own individuality of taste,—he knows altogether too much. This is the Scotch phase of knowing too much; but there is an American one that is even worse, and which puts a raw edge upon country socialities.

I find no man so disagreeable to meet with, as one who knows everything. Of course we expect it in newspaper editors, and allow for it. But, to meet a man engaged in innocent occupations—over your fence, who is armed cap-a-pie against all new ideas,—who 'knew it afore,' or 'has heerd so,' or doubts it, or replies to your most truthful sally 't'ain't so, nuther,' is aggravating in the extreme.

There is many a small farmer, scattered up and down in New England, whose chief difficulty is—that he knows too much. I do not think a single charge against him could cover more ground, or cover it better. It is hard to make intelligible to a third party, his apparent inaccessibility to new ideas, his satisfied quietude, his invincible *inertium*, his stolid, and yet shrewd capacity to resist novelties, his self-assurance, his scrutinizing contempt for outsidedness of whatever sort—his supreme and ineradicable faith in his own peculiar doctrine, whether of politics, religion, ethnology, ham-curing, manuring, or farming generally.

It is not alone that men of this class cling by a particular method of culture, because their neighborhood has followed the same for years, and the results are fair; but it is their pure contempt for being taught; their undervaluation of what they do not know, as not worth knowing; their conviction that their schooling, their faith, their principles, and their understanding are among God's best works; and that other peoples' schooling, faith, principles, and views of truth—whether human or Divine—are inferior and unimportant.

Yet withal, there is a shrewdness about them which forces upon you the conviction that they do not so much dislike to be taught, as dislike to *seem* to be taught. They like to impress you with the notion that what you may tell them is only a new statement of what they know already. It is inconceivable that anything really worth knowing has not come within the range of their opportunities; or if not theirs, then of their accredited teachers, the town school-master, the parson, the doctor, or the newspaper. In short, all that they do not know which may be worth knowing, is known in their town, and they are in some sort partners to it.

Talk to a small farmer of this class about Mechi, or Lawes, or the new theory of Liebig, and he gives a complacent, inexorable grin—as much as to say—

"Can't come that stuff over me; I'm too old a bird."

So indeed he is; and a tough bird at that. His mind is a rare psychological study; so balanced on so fine a point, so immovable,—with such guys of prejudice staying him on every side,—so subtle and yet so narrow,—so shrewd and yet so small,—so intelligent and yet so short-sighted If such men could bring themselves to think they knew less, I think they would farm far better.

Opportunity for Culture.

THERE is a plentiful crop of orators for all the agricultural fairs (most of them city lawyers, not knowing a Devon from a Hereford), who delight in expatiating upon the opportunities for culture afforded by the quiet and serenity of a farm-life. Now there is no life in the world, which, well husbanded, has not its opportunities for culture; but to say that the working-farmer's life is specially favored in this respect, is the grossest kind of an untruth.

——Long evenings, forsooth! And the orator who talks in this style is probably crawling out of his bed at eight in the morning, while the farmer is a-field since four. And are not these four hours to be made good to him in sleep or rest? The man who rises at four, and works all day, as farmers work, or who is even a-field all day, is sleepy at nine P. M. It is not, perhaps, a graceful truth ; but it is a physiological one. Nothing provokes appetite for sleep so much, as out-of-door life. You may over-strain the nervous system, and dodge the night ; but a strain upon the muscular system must have its balance of repose. There are, indeed, exceptional cases, where a working man with an undue preponderance of brain, will steal hours between his labor for intellectual cultivation ; but he does it under difficulties, which he is the first to recognize and deplore. Even the most skilled of working farmers arrive at their conclusions by an intuitive sagacity, which is wholly remote from the logical processes of books ; and their straight-forward common-sense, however correct in its judgments, grows into a distaste for the subtle arts of rhetoric.

During the more leisure period of winter, the practical mind of the farmer will gravitate more easily toward mechanical employments, than toward those which are intellectual. He will have his Agricultural-journal and others, may be, to whose reading he will bring a ripe and hardy judgment. But his thoughts will be more among his cattle and his bins, than among books. "He cannot get wisdom that glorieth in the goad, and that driveth oxen."

There may be a spice of exaggeration in the dogma of Ecclesiasticus; but whoever undertakes the profession of working-farmer, must accept its fatigues and engrossments, and honor them as he can. It is a business that will not be halved. Vulcan can make no Ganymede—strain as he will. The horny hands, the tired body, the hay-dust and the scent of the stables are inevitable. The fine young fellow, flush with Johnston's Elements, and buoyant with Thomson's Seasons, may rebel at this view of the case; but let him take three hours in a hay-field of August—behind a revolver (rake), with the reins over his neck, the land being lumpy, and the colt dipping a foot over the traces at the end of every bout, and I think he will have sweaty confirmation of its general truth. Or let him try a day at the tail of a Michigan-plough, in a wiry and dusty last-year's stubble:—the horses are fresh and well trained, and the plough enters bravely to its work—smoothly at first, but presently an ugly stone flings it cleanly from the furrow, and there is a backing,—a heavy tug, and on he goes with his mind all centred in the plough-beam, and nervously watching its little pitches and yaws; he lifts a hand cautiously to wipe the perspiration from his forehead (a great imprudence), and the plough sheers over gracefully, and is out once more. There is a new backing and straining, and the plough is again in place; no more wiping of the forehead until the headlands are reached. Watery blisters are rising fast on his hands, and a pebble in his shoe is pressing fearfully on a bunion; but at the headland he finds temporary relief, and a small can of weak barley-water. Refreshed by this, but somewhat shaky in the legs, he pushes on with zeal—possibly thinking of Burns, and how he walked in glory and in joy,

> "Behind his plough,
> Upon the mountain side,"

—and wondering if he really did? There are no 'wee-tipped' daisies to beguile him; not a mouse is stirring; only a pestilent mosquito is twanging somewhere behind his left ear, and a fine aromatic powder rises from the dusty stubble and tickles his nostrils. So he comes to the headland once more and the can; if he had a copy of Burns in his pocket, it might be pleasant for the fine young fellow to lie off under the shade for a while, and 'improve his mind.' But he has no Burns—in fact, no pocket in his overalls; besides which, the season is getting late; he must finish his acre of ploughing. Over and over he eyes the sun—it is very slow of getting to its height, and when noon comes it finds him in a very draggled and wilty state; but he mounts one of the horses, and the mate clattering after, he leads off to the barn and the bait ing. He has a sharp appetite for the beef and the greens, but not much, at the nooning, for Burns or Bishop Butler. The return to the field haunts him; but the work is only half done. Rubbing his puffy hands with a raw onion (by the advice of Pat), he enters bravely upon a new bout of the ploughing. The sun is even more searching than in the morning; the mosquitoes have come in flocks; the bunion, aggravated by the morning's pebble, angers him sorely, and destroys all his confidence in the commentators upon Burns.

At night, more draggled and wilted than at noon, he turns out his team, and if he means systematic farm-work, will give the horses a thorough rubbing-down; afterward, if he cherish cleanly prejudices,—the fine young fellow will have need for a rubbing-down of himself. This refreshes, and gives courage for the milking—which, with those puffy fingers, is no way amusing. Again the appetite is good—even for a cut of salt-beef, and dish of cold greens. Thereupon Pat, the Irish lad, sits upon the doorstep and ruminates,—with a short, black pipe in his mouth. Our draggled young friend aims at something better; it is wearily done; but at least the show shall be made. The candle is lighted, and a book pulled down—possibly Prof. Johnson on Peats; the millers dart into the flame; peats, and hydrates, and oxides, and peats again, mix strangely; a horned beetle dashes at his forehead, and makes him wakeful for a moment; there is a frog droning in the near pond very drowsily—'peats—peats—peats;' the drift of the professor is lost; Pat ruminates on the step; a big miller flaps out the flame of his candle;—it is no matter—our fine young fellow is in a sound snooze.

So much for the working farmer; and we cannot have armies without privates; and privates are many of them 'fine young fellows.'

Isolation of Farmers.

I AM reminded that a farmer has no need to fag himself with hard field work. To a certain extent this is true; but only "A master's eye fattens the horse, and only a master's foot the ground."

If farming be undertaken as an amusement, absence is possible; indeed, the longer the absence, the greater the amusement—to the onlookers; but if farming be undertaken as a business, presence is imperative—presence, with its associations, and its comparative isolation.

Of the more familiar associations, a type may be had in Pat, sitting on the doorstep at dusk, ruminating and smoking a black-stemmed pipe. The isolation is less obvious, but more galling. Farms do not lie extensively in cities; and the least fear we live under,—is one of mobs. In fact, there is not even a habit of congregation in farmers. They meet behind the church, between services,—in a starched way; they drive to town-meetings in their best toggery, and discuss ballotings and the weather—possibly linger an hour or two about the tavern or a pet grocer's; but they do not meet as townspeople meet—on the walk, over counters, on the railway, in the omnibus, and in each other's houses. I have already taken occasion to dust out their darkened parlors; but the dust will gather again. They have no Market-Fairs* which will bring them together with samples of their crops, to compare notes, and prices, and methods of culture.

There is no coherence of the farmers as a body—no trade-guild—no banding of endeavor to work a common triumph, or to ferret out a common abuse. For years, in many parts of New England, the sheep culture has been entirely ruined by the ravages of lawless town-dogs; and the farmers groan over it, and bury the dead sheep, and whisper valorously between church services about bludgeons and buck-

* A strong effort, I am glad to see, is making to establish them in various parts of the country. In my own neighborhood the old town of Cheshire has made a bold stride in this direction, and I trust not in vain. They are worth more to the true interests of farming than all the horse-trotting fairs which could be packed into a season.

shot, but never make a concerted urgent protest; or if they rally so far as to send one of their own people to the Legislature,—he, poor fellow, does not pass ten days under the fingers of the lobbyists, but he sinks into the veriest dribblet of a politician; and gives the last proof of it, by making a pompous speech on 'Federal Relations,'—not worth the carcass of a ewe lamb.

Under these conditions, any new and valuable methods of farm-practice do not spread with any rapidity; they hobble lamely over innumerable flanking walls. It is possible they may get an airing in the Agricultural journals; but good and serviceable as these journals are, their statements do not influence, like personal communications. Reforms want the ring of spoken words, and some electric social chain traversing a whole district, and flashing with neighborly talk.

The man of education, giving himself over to the retirement of a farm-life, will find this isolation, sooner or later, grating sorely. Whatever love of the pursuit—its cares, indulgences, attractions, successes—may engross him, a certain attrition with the world is as necessary to his mental health, and briskness of thought—as a rubbing-post for his pigs. He may let himself off in newspapers, or he may thumb his library and the journals, but these offer but dead contact, and possess none of that kindling magnetism which comes from personal intercourse. Type grows wearisome at last, however stocked with information and gorgeous fancies; and a man frets for the lively rebound of discussion.

Friends from the city may drop upon you from time to time, exercising this compassion for your retirement; and they treat you compassionately. Of course the novelty of the scene and the life has charms for any metropolitan, whatever his tastes; and he bears himself very briskly at the first. The view is charming; the well-water is charming; the big oaks (they are all maples) are charming. And his eye falls upon a riotous hedge of Osage-orange, "Dear me, that's the hawthorn; how beautiful it is!"

Of course you do not correct him; in fact, you partake of his exhilaration, and seem to see things with new eyes.

"And, bless me, here's your boy (its a girl); how old is he?" (patting her head).

What a fine flow of spirits he is in, to be sure! You show him up and down your grounds (always 'your grounds,' he calls them, if it be only a potato garden).

Presently his eye lights upon a blooming Weigelia. "Ah, a dwarf apple! and do you go largely into fruit?" upon which you offer him a Red-Astrachan, and remark that the Weigelia has not borne thus far; it is a Chinese shrub, and little understood as yet.

"Is it possible—Chinese! so far;—it seems to thrive." And it does.

And you stroll with him upon the hill; though you cannot but see that his mind is warping back to 'laryngal affections,' or 'half-of-one-per-cent. off.'

A lucky interruption appears, in the shape of a fine Devon cow. You venture to call his attention to her, and ask if she is not a fine animal?

"Admirable!" and with a kind interest, he asks —if she isn't a short-horn?

"Not a Short-horn," you reply; and in way of apology for his error, remark that she has broken off one of her horns in the fence.

At which he says,—"Ah, I see now;—but resembles the Short-horns, doesn't she?"

"Yes—" you return, mildly—"a little; her legs are like; and I think she carries her tail—a good deal in the Short-horn way."

At which he is himself again, and is prepared for a new farm venture. It comes presently, as a fine brood of Bremen geese waddle into sight.

"Muscovies?"

"No, not ducks—geese—Bremen geese, but resemble the Muscovies;" (as unlike as they are to sea-fowl; but shall not a host keep his guest in good humor?)

"I shouldn't have known 'em from Muscovies," he says. And I really don't suppose that he would.

A good-natured city-guest, who comes to see you in your retirement, is very apt to talk in this strain upon farming matters. It is engaging, but not improving.

You stroll, by and by, into the library; and leave him for a few moments lounging in the arm-chair, while you slip out to give some orders to the ditchers in the meadow.

Upon your return, entering somewhat brusquely (expecting to find him deep in some book), you waken him out of a sound sleep.

"Upon my word," he says, "this is a beautiful air; if I lived here I should sleep half my time."

The reflection is a somewhat dismal one,—though well meant.

All this, however, illustrates what I want to say —that the citizen engrossed in active professional or business pursuits, when he visits a farm friend, goes with the very sensible purpose and hope—of escaping for a while the interminable mental strain of the city, and of giving himself up to full relaxation. And this fact makes the isolation of which I have spoken, more apparent than ever.

And it is an isolation that cannot altogether be left behind one. On your visits to the city, friends will remark your seediness, not unkindly, but with an oblique eye-cast up and down your figure—as a jockey measures a stiff-limbed horse—long out to pasture. You may wear what toggery you will—keeping by the old tailors, and showing yourself *bien ganté*, and carefully read up to the latest dates; still you shall betray yourself in some old dinner-joke—dead long ago. And the friends will say kindly, after you are gone, "How confoundedly seedy Rus. has grown!"

Were this all, it were little. But the clash and alarum of cities have stirred things to their marrow, which you know only outsidedly. The great nervous sensorium of a continent,—with its wiry nerves raying like a spider's web, in all directions,—is packed with subtle and various meanings, which you, living on an outer strand of the web, can neither understand nor interpret. Mere accidental contact will not establish affinity. In a dozen quarters a boy puts you right; and some girl tells you newnesses you never suspected. The rust is on your sword; thwack as hard as you may, you cannot flesh it, as when it had every day scouring into brightness.

Dickering.

SOMETIME or other, if a man enter upon farm life—and it holds true in almost every kind of life—there will come to him a necessity for bargaining. It is a part of the curse, I think, entailed upon mankind, at the expulsion from Eden,—that they should sweat at a bargain. When a Frenchwoman with her hand full of gloves,—behind her dainty counter,—asks the double of what her goods are worth, you are noway surprised. You accept the enormity, as a symptom of the depravity of her race,—which is balanced by the suavity of her manner.

But when a hard-faced, upright, sabbath-keeping New-England bank-officer or select-man, asks you the double, or offers you the half, of what a thing is really worth, there is a revulsion of feeling, which no charm in his manner can drive away. Unlike the case of the French shop-woman, I feel like passing him—on the other side of the street.

And yet all this is to be met (and conquered, I suppose) by whoever has butter, or eggs, or hay, or fat-cattle to sell. I ventured once to express my surprise to a shrewd foreman who had charge of this business—for I manage it by proxy as much as I can—that a staid gentleman with his ten thousand a year of income, should have insisted upon a deduction of two cents a bushel in the price of his potatoes, in view of a quart of small ones, that had insinuated themselves in the interstices: I think I hear his horse-laugh now, as he replied—"Why, sir, it's the way he grew rich."

The idea struck me as novel; but upon reflection I am inclined to think it was well based. As I said, —often as possible, I accomplish this business by proxy; and, in consequence, have made some bad debts by proxy. But proxy is not always available. There are customers who insist upon chaffering with the 'boss.' Such an one has dropped in, on a morning in which you happen to be deeply engaged. He wishes to 'take a look' at a horse, which he has seen advertised for sale. The stable is free to his observation, and the attentive Pat is at hand; but the customer wants a talk with the 'Squire.'

It is a staunch Canadian horse, for which you have no further use. You paid for him, six months gone, a hundred and fifty dollars, and you now name a hundred dollars as his price. I never yet met a man who sold a horse for as much as he gave—unless he were a jockey; I never expect to.

"Mornin', Squire."

"Good morning."

"Bin a lookin' at y'er hoss."

"Ah!"

"Middlin' lump of a hoss."

"Yes, a nice horse."

"D'n know as you know it, but sich hosses an't so salable as they was a spell back."

"Ah!"

"They're gittin' a fancy for bigger hosses."

Silence.

"Put that pony to a heavy cart, and he wouldn't do nothin'."

"You are mistaken; he's a capital cart-horse."

"Well, I don't say but what he'd be handy with a lightish load. Don't call him spavined, do ye?"

"No, perfectly sound."

"That looks kinder like a spavin"—rubbing his off hind leg.

"An't much of a hoss doctor, be ye?"

"Not much."

"Don't kick, dooz he?"

"No."

"Them little Kanucks is apt to kick."

Silence, and an impatient movement, which I work off by pulling out my watch.

"What time o' day 's got to be?"

"Eleven."

"Thunder! I must be a goin';—should like to trade, Squire, but I guess we can't agree. I s'pose you'd be askin' as much as—sixty—or—seventy dollars for that are hoss—wouldn't ye?"

"A hundred dollars is the price, and I gave fifty more."

"Don't say! Gave a thundering sight too much, Squire."

"Pat, you may put up the horse; I don't think the gentleman wants him."

"Look o' here, Squire;—ef you was to say—something—like—seventy, or—seventy-five dollars, now,—there might be some use in talkin'."

"Not one bit of use," (impatiently)—turning on my heel.

"——Say, Squire,—ever had him to a plough?"

"Yes."

"Work well?"

"Perfectly well."

"Fractious any? Them Kanucks is contrary critters when they've a mind to be."

"He is quite gentle."

"That's a good p'int; but them that's worked till they git quiet, kinder gits the spirit lost out on 'em—an't so brisk when you put 'em to a waggin. Don't you find it so, Squire?"

"Not at all."

"How old, Squire, did ye say he was?" (looking in his mouth again).

"Seven."

"Well—I guess he is; a good many figgers nigher that, than he is to tew—any way."

"Patrick, you had better put this horse up."

"Hold on, Squire," and taking out his purse, he counts out—"seventy—eighty,—and a five,—and two,

—and a fifty—there, Squire, 'tant worth talkin' about; I'll split the difference with ye, and take the hoss."

"Patrick, put him up."

At which the customer is puzzled, hesitates, and the horse is entering the stable again, when he breaks out explosively—

"——Well, Squire, here's your money; but you're the most thunderin' oneasy man for a dicker that I ever traded with—I'll say that for ye."

And the horse is transferred to his keeping.

"S'pose you throw in the halter and blanket, Squire, don't ye?"

"Give him the halter and blanket, Patrick."

"And, Patrick, you 'ant nary old curry-comb you don't use, you could let me have?"

"Give him a curry-comb, Pat."

"Squire, you're a *clever* man. Got most through y'r hayin'?"

"Nearly."

"Well, I'm glad on't. Had kinder ketchin' weather up our way."

And with this return to general and polite conversation, the bargaining is over. It may be amusing, but it is not inspiriting or elevating. Yet very much of the country-trade is full of this miserable chaffering. If I have a few acres of woodland to sell, the purchaser spends an hour in impressing upon me his 'idee'—that it is scattered and mangy, and has been pirated upon, and that wood is 'dull,' with no prospect of its rising; if it is a cow that I venture in the market, the proposed purchaser is equally voluble in descriptive epithets, far from complimentary; she is 'pooty well on in years,' rather scrawny, 'not much for a bag,'—and this, although she may be the identical Devon of my Short-horn friend. If it is a pig that I would convert into greenbacks—he is 'flabby,' 'scruffy,'—his 'pork will waste in bilin'.' In short if I were to take the opinions of my excellent friends the purchasers—for truth, I should be painfully conscious of having possessed the most mangy hogs, the most aged cows, the scrubbiest veal, and the most diseased and stunted growth of chestnuts and oaks, with which a country-liver was ever afflicted.

For a time, in the early period of my novitiate, I was not a little disturbed by these damaging statements; but have been relieved on learning, by farther experience, that the urgence of such lively falsehoods is only an ingenious mercenary device for the sharpening of a bargain. But while this knowledge puts me in good temper again with my own possessions, it sadly weakens my respect for humanity.

Amateur farmers are fine subjects for these chafferers; they yield to them without serious struggle

The extent and the manner of their losses, under the engineering abilities of those wiry old gentlemen who drive sharp bargains, is something quite beyond their comprehension. It would be well if harm stopped here. But this huckstering spirit is very leprous to character. It bestializes;—it breaks down the trader's own respect for himself, as much as ours. The man who will school himself into the adoption of all manner of disguisements about the cow he has to sell, will adopt the same artifices and quibbles about the opinions he wishes to force upon your acceptance. Let him mend by showing all the spavins in the next horse he has for sale (there will be some, or he would never sell); and his reformation is not altogether hopeless.

The Bright Side.

THUS far I have been dealing with the shadows—heavily laid on; let me now, with a finer brush, touch in the lights upon my picture. The chemical puzzles, the disappointments, the isolation, the fatigues, the chaffering bargainers do not fully describe or give limit to the good old profession of farming. And even when these clouds—hindrances I call them—most accumulate, the kindly sun flashes through, warming all the fields below me into golden green, and a kindly air stirs all the poplars into silver plumes, and I am beguiled into a new and a more admiring estimate of the country life.

Arcadia with its sylvan glories comes, drifting to my vision, and the pleasant Elian fields sloping to the sea. A stately Greek gentleman—Xenophon—who has won great renown by his conduct of an army among the fastnesses of Armenia, and on the borders of the Caspian, has retired to his estates on the Ionian waters, and writes there a book of maxims for farm management, which are not without their significance and value to every farmer to-day. And hitherward, across the blue wash of the Adriatic, in the midst of the Sabine country, which is northward and eastward of Rome, I know a Roman farmer—Cato—who has been listened to with rapt attention in the Roman Senate, and who—centuries before the time when Horace was amateur agriculturist, and planted Soracte and Lucretilis in his poems—wrote so minutely, and with such rare sagacity, upon all that relates to country living, and to country thrift, that I might to-morrow, in virtue of his instructions only, plant my bed of asparagus, and so dress and treat it (always in pursuance of his directions) as to insure me for the product a prize at the County-Fair ;—if, indeed, the shoots did not rival those famous ones of Ravenna—of which Pliny speaks—weighing three to the pound.

I know a poet too, whose music floating over Italy, before yet the battle blasts of her direst civil strife were done, weaned soldiers from their blood scent to the tranquil offices of husbandry; and that melody of the Georgics is floating still under all the ceilings of all the school-houses of New England. The most pretentious and the most ambitious of the later emperors of the West—Porphyrogenitus—has left no more enduring monument of his reign, than the compend of agricultural instructions, compiled under his order, and bearing title of "Geoponica Geoponicorum."

I observe, too, in my card-basket, the address of a certain Pietro di Crescenzi, who has come all the way from the fourteenth-century-Bologna to pay me a visit—in a tight little surtout of white vellum that smacks of the loves of Bembo, or of the wickedness of the Borgia; and who has talked of horses and cattle, and wheat-growing, and vegetable-raising, as familiarly as if he were justice of the peace in our town. Lord Bacon has contributed to our stock of information about garden culture, and the elegant pen of Lord Kames has illustrated the whole subject of practical husbandry. But I do not cite these names for the sake of making any idle boast of the antiquity and dignity of the craft; we have too much of that, I think, in our agricultural addresses.

We live in days when a calling—whatever it may be—cannot find establishment of its value or worth, in the echoes—however resonant and grateful—of what has once belonged to it, or of the dead voices that honored it. The charms of Virgil and the shrewd observations of Cato will go but a little way to recommend a country life in our time, except that life have charms in itself to pique a man's poetic sensibilities—and lessons in every field and season, to tempt and reward his closest observation.

Yet it is very remarkable how nearly these old authorities have approached the best points of modern practice; and again and again we are startled out of our vanities by the soundness of their suggestions. Rotation of crops, surface drainage, ridging of lands, composting of manures, irrigation, and the paring and burning of stubble-lands are all hinted, if not absolutely advised, in treatises written ten centuries ago. Nor have I a doubt but that a shrewd man acting upon the best advices which are to be found in the various books of the Geoponica (the latest not later than the sixth century), and with no other instructions whatever—save what regards the dexterous use of implements—would manage a grain field, a meadow, or an orchard, better than the half of New England farmers.

At first blush, it seems very discouraging to think

that we have put no wider gap between ourselves and those twilight times. The gap is, however, far wider than it seems; for while those old gentlemen made good hits in their practice, they rarely announced a principle on which good cultivation depended, but they were egregiously at fault. The centuries, with their science and added experience, have solved the *reasons* of things; not all of them, indeed—as Liebig in his last book needlessly tells us—but enough of them to enlist a more intelligent method of culture. The ancients recommended a rule of practice, because it had succeeded in a score or a hundred of trials; but if some day it failed, they must have groped considerably in the dark for a cause. *We* lay down a rule of practice in obedience to certain clearly determined natural laws; and if failure meets us, we know it is due—not to falsity of the laws—but to some one of a rather wide circle of contingencies, not foreseen or provided against. And it is the due adjustment and measurement of precisely this circle of contingencies—whether belonging to weeds, weather, or markets—which most thoroughly tests the sagacity of the modern farmer.

This sagacity is of far larger service, than I think scientific farmers are willing to admit. Over and over it happens that some uncouth, raw, strapping, unread man succeeds, year after year, in making crops which astonish the neighborhood. You know he has no science,—nitrogen is Greek to him; sulphuric acid, for all he can tell, might lie in the juice of an apple; he knows nothing of fermentations—nothing of physiology, yet his crops are monstrous. His tools are something old, though firm and compact; his team is always in good order, although his barns may be somewhat shaky.

He could not himself explain to you his success; you perceive that he manures well, that he ploughs thoroughly, that he plants good seed, that he hoes in season. This is all; but all is so well timed by a native sagacity—by an instinctive sense (as would seem) of the wants and habits of the crop, growing out of close observation—that the success is splendid. A man sets up beside him, and buys guano and fish, and the best tools, and employs a chemist to analyze his soil—but his crops do not compare with those of his rude neighbor, who sneers at chemistry and fine farming. Of course I do not mean to join him in his sneers; I only mean to illustrate how a large sagacity, guided by its own instincts, has very much to do with good farming; and in a way not clearly explicable—certainly not explicable by its possessor.

Just so, you will sometimes find, far back in the country, a shrewd old physician, utterly unread in the new books, who laughs at the *Gazette des Hôpitaux* and the Chirurgical, and yet who has that rare insight which enables him to detect and wrestle with disease strangely well. His long observation, his comparison of trifles, his estimate of the moral forces at work are so just and discriminating, that he brings a tremendous power of judgment to the case. Put him in a room for consultation, and his gray eye tweaks, his lips work nervously; he cannot enter into the learned discourse of the younger men of the profession; he is dazed by it all—wishing he were learned, if learning helps; but when appeal is made to him, there is such clear, sagacious, homely cut-down into the very marrow of the difficulty, as absolutely confounds the young doctors; all this, not because he does not carry learning, but because he carries brain—and uses it.

Any man with good brains may succeed in farming—if he uses them. By this, I mean that any man with a clear head—though not specially crammed with information—and who brings a cool, sagacious, unblinking outlook to the offices of husbandry, will succeed, without a knowledge of the principles on which its more important operations are based. And the practice of such a man, if faithfully recorded in all its details, would be of more service in the illustration of scientific laws, than the halting experience of a half dozen neophytes, who work by the vague outline of some pet theory. I had rather have such a man for tenant, than one fresh from the schools, bringing an exaggerated notion of salts, and a large contempt for sagacity. If on some day of latter summer the milch cows rapidly fall away in their 'yield,' I should expect the latter to puzzle himself about the sudden exhaustion of some particular constituent of the milk food, and to multiply experiments with bran or bone meal for its supply; but I should expect the sagacious veteran, under the same circumstances, with a bold philosophy, to attribute the shortcoming to the scorching suns of August, that have drunk up all the juices of the grass; and I should expect him to meet the want by a lush and succulent patch of pasturage, which his foresight has kept in reserve.

Business Tact.

AKIN to this sagacity is a certain business tact, which is a large helper to whoever would successfully engage in agricultural pursuits. It implies and demands adaptation of crops to soils, exposure, and the market wants. It is eminently opposed to the drowsiness in which a good many honest country-livers are apt to indulge. It reckons time at its full value; it does not lean long on a hoe-handle for gossip.

The farmer who turns his capital very slowly, and only once in the year, is not apt to be quickened into business ways and methods. The retired trader, who plants himself some day beside him, bringing his old prompt habits of the counter, will very likely, if a shrewd observer, outmatch him in a corn crop,—outmatch him in pork,—outmatch him in everything, if the year's balance were struck and shown. And all this in spite of the trader's comparative inexperience, and by reason only of his superior business tact.

The finest shows of fruits at the autumn fairs—excepting always those of the professed nurserymen—are made, in three cases out of five, by mechanics, or by business men, who have brought to this little episode in their life, the methodical habits, and the observance of details, which govern their ordinary business duties. Not being in the way of leaving book accounts, or stock on hand, to take care of themselves, they are no more inclined to leave an investment in trees or orcharding—to take care of itself. They reckon upon care at the outset, and they bestow it. The farmer, who has complacently smiled at their inexpèrience in tillage, and is confounded by the results, will loosely attribute them all to a lavish and thriftless expenditure of money. But the conclusion is neither logical, nor warranted,—in the majority of instances,—by the facts. No superior fruit can be grown without labor and extreme care, and if these be controlled by a business system, they will be far more economically bestowed, than when subject to no order in their application.

From time to time I observe that some venerable old gentleman in my neighborhood is overtaken by one of those sporadic fevers of improvement, which will sometimes, and very strangely, attack the most tranquil and self-satisfied of men. The attack is a slight one, of the orchard type. He consults far and near in regard to the best sorts of fruit. He devotes to the experiment one of his best lots, reserving the very best for his next year's patch of potatoes. The land he reckons in 'good heart,' since he has just taken off a heavy crop of corn. He digs his holes, after an elaborate system of garden measurement and stake-driving, which, to his poor, fagged brain, seems the very climax of geometric endeavor. The young trees are carefully staked, and for a year or two show a thrifty look. But the spring temptation to put a crop between the roots is irresistible; the ploughing oxen browse a few—knock over a few—break off a few. This maddens our friend into a 'laying-down' of the orchard to grass; he half promises himself, indeed, that he will give hand-cultivation to the trees,—but he does not; his fever is abating, and so is his orcharding. The mosses fasten on the young trees, the borers play havoc, the caterpillars strip them, the rank grass strangles them.

From beginning to end there has been no business forecast of the requisite labor involved, no method in its prosecution—no estimate of the scheme as a business operation.

It is certain that by a special dispensation of Providence in favor of those who make up the bulk of the human family, a man may secure a simple livelihood in agricultural pursuits, with less of energy, less of promptitude, less of calculation, and greater unthrift generally, than would be compatible with even this scanty aim, in any other calling of life. With a respectable crop insured by only a moderate amount of attention and activity, the temptation to a lazy indifference, and a sleepy passivity, is immense. There are farmers who yield to the temptation gracefully and completely. The stir, the wakefulness, the promptitude that seize upon new issues, develop new enterprises, create new demands, are as foreign to the majority of landholders, as a ringing discussion of new topics, or a juicy haunch of Southdown, to their tables.

But whatever may be the triumphs of business tact—and of a just apportionment of capital, between land and implements, or fertilizers, the real question with a man of any considerable degree of cultivation who meditates country life,—is not whether legitimate attention will secure a tolerable balance sheet, and the fattening of fine beeves, but whether the life and the rural occupations offer verge and scope for the development of his culture—whether land and landscape will ripen under assiduous care into graces that will keep his attachment strong, and enlist the activities of his thought?

Let us inquire.

Place for Science.

BECAUSE a man cannot revolutionize farming and its practice by clean copies of Boussingault and Liebig under his arm, or upon his table, it by no means follows that an intelligent person who is concerned in rural occupations may not profitably give days and nights to their study. Because we cannot conquer all diseases, and clearly explain all the issues of life and death by the best of medical theories, it by no means follows that the best medical practitioner should therefore abandon all the literature of the subject. The scientific inquirers who direct their view to agricultural interests, deal with problems which are within the farmer's domain; and if their solutions are not always final or directly available, the very intricacy of their nature must pique his wonder, and enlist his earnest inquiry.

A magnificent mystery is lying under these green coverlets of the fields, and within every unfolding germ of the plants. The chemist is seeking to unriddle it in his way; while we farmers,—by grosser methods,—are unriddling it, in ours. Checks and hindrances meet us both; both need an intimate comparison of results for progress. If we sneer at the chemist for his shifting theories in regard to the nitrogenized manures—no one of which is sufficiently established for the direction of a fixed practice—the chemist may return the sneer with interest, when he sees us making such application of a valuable salt, as shall lock up its solubility and utterly annul its efficacy. It is a pretty little duel for our intelligent observer to watch: the chemist fulminating his doctrines, based on formulas and an infinity of retorts; and we, replying only with the retort—courteous and practical. But always the unfathomable mystery of growth—vegetable and animal—remains; the chemist seeking to explain it, and we only to promote it. If the chemist could explain by promoting it, he would turn farmer; and if farmers could promote it by trying to explain it, they would all turn chemists.

Many good people, of a short range of inquiry, and a shorter range of reflection, imagine that when the agriculturist has, by the chemist's aid, determined the elements of his crops, and by the same aid, determined the merits of different bags of phosphates or guanos, that nothing remains but to match these chemical colors as he would match colts,—and the race is won. They fancy that the new analyses and experiments—so delicate and so elaborate—are by their revelations reducing the art of farming to a simple affair of the mechanical adjustment of regularly-billeted chemical forces. There could not be a greater mistake made; so far from simplifying, the new investigations demand a larger practical skill, since the conditions under which it works are amplified and extended. The old bases of procedure, if faulty, were at least compact; the experimental farmer dealt with but few, and those clearly defined; but scientific investigation, by its refining processes, has split the old bases of action into a hundred lesser truths, each one of which must be taken into the account, and modify our operations.

There was a time, for instance, when science, observing that a living plant built itself out of the debris of dead plants, declared for the primal necessity of a large supply of decayed vegetable material. This at least was simple, and the farmer, if he had only his stock of *humus*, left the further fulfilment of the miracle of growth to wind and weather. In process of time, however, science detected the rare luxuriance which ammonia imparts to plant foliage, and after refining upon the observation, declared for *nitrogen* as the great needed element; schedules were prepared and widely published, in which the various manures were graduated in value, in strict accordance with their respective admixtures of nitrogenous material. The quiet farmer accepts the theory, and considers the wonderful effects that follow the application of the droppings from his dovecot, a demonstration of its truth.

But he has hardly nestled himself warmly into this belief,—modified to a degree by the *humus* doctrine, than a distinguished chemist comes down upon us all with the representation—supported by a large array of figures—that nitrogen is already present in ample quantity in almost all soils, and that the vital necessity in the way of fertilizers, is the mineral element of the plant. This splinters once again the compactness of our purpose, and puts us upon a keen scent for the soluble phosphates; though without destroying our faith in good vegetable-mould and strong-smelling manures.

And not only in this direction, but also in what relates to the feeding of animals, the germination of seeds, the comminution of soils, the chemical effects of air, and light, and warmth—we have a hundred minute truths by which to adjust our practical management, where we had formerly less than a score of gross ones. And in this adjustment—modified still further by a great many physiological and meteorological considerations—I think a man of tolerable parts might find enough to lay his mind to very closely, and to encourage some activity of thought.

There will be disappointments—as in every sphere of life. I have felt them keenly and often. The *humus* has baffled my expectations, and my potatoes; the nitrogenous riches have shot up into thickets of rank and watery luxuriance; the phosphoric acid has oozed into some unthrifty combination, or has remained locked up in an unyielding nugget of Sombrero. But little disappointments count for nothing, when (as now) we are reckoning the pabulum which agricultural employments furnish for intellectual activity. The rural adventurer may not only regale himself with a considerable series of nice chemical puzzles at every cropping-time, but he may give his thoughts to original investigation of the habits of the plants themselves; the career of a Decandolle could have had no finer start-point than a country farm with its living herbaria, and its opportunities for observation; we want a good monograph of our great national crop of maize—so soon as the man shall appear to make it. We want, too, some Buffon (without his foppery) to unearth our field mice, and to put a great tribe of insect depredators to flight.

Æsthetics of the Business.

WHAT is needed, perhaps more than all else, in our agricultural regions, is—such intelligible, imitable, and economic demonstrations of the laws of good taste, as shall provoke emulation, and redeem the small farmer—unwittingly, it may be—from his slovenly barbarities and his grossness of life. Here is verge, surely, for a man's cultivation, for his aptitude, and for those graces which shall fix his attachment while they plead their lessons of appeal.

It seems hardly necessary to urge a necessity for this direction of effort. There is certainly no race of country-livers in the world, who, with equal, or even a kindred intelligence, are so destitute of all sense of the graces of life and home, as the small New England farmers.

A certain stark neatness, confined mostly to kitchens, pantries, and such portions of the door-yard as are under the eye of the goodwife, mostly limits their efforts in this direction. It may be that a staring coat of white paint upon the house completes the investiture of charms; while, at every hand, heaps of rubbish—cumbering the public road—and piles of straggling wood, dissipate any illusion which a well-scrubbed interior, or the fresh paint, may have created.

Here and there we come upon a certain neatness and order in enclosures, buildings, and fields; but ten to one the keeping of the picture is absolutely ruined by the slatternly condition of the highway, to which,—though it pass within ten feet of his door,—the farmer, by a strange inconsequence, pays no manner of heed. He makes it the receptacle of all waste material, and foists upon the public the offal, which he will not tolerate within the limits of his enclosure. And the highway purveyors are mostly as brutally unobservant of neatness as the farmer himself; nay, they seem to put an officious pride into the unseemliness and rawness of their work; and it is only by most persistent watchfulness that I have been able to prevent some bullet-headed road-mender from digging into the turf-slopes at my very door.

Here and there I see, up and down the country, frequent attempts at what is counted ornamentation—fantastic trellises cut out of whitened planks, cumbrous balustrades, with a multitude of shapeless finials, or whimsical pagodas—imitations of what cannot be imitated, even if worthy;—but of the hundred nameless graces, wrought of home material, delighting you by their unexpectedness, piquing you by their simplicity, and winning upon every passer-by, by their thorough agreement with landscape, and surroundings, and the offices of the farmer, I see far less. The only idea of elegance and beauty which finds footing, is of something extraneous—outside his life—not mating with his opportunities or purposes—and only to be compassed, as a special extravagance, upon which some town joiner must lavish his 'ogees,' and which shall serve as a blatant type of the farmer's 'forehandedness.' This is all very pitiful; it gives no charm; it educates to no sense of the tender graces of those simple, honest adornments which ought to refine the country-liver, and to refine the tastes of his children. I am not writing in any spirit of sentimental romanticism. If Arcadia and its pastorals have gone by (and I think they have), God, and nature, and sunshine, have not gone by. Nor yet the trees, and the flowers, or green turf, or a thousand kindred charms, which the humblest farmer has in his keeping, and may spend around his door and homestead, with such simple grace, such affluence, such economy of labor, such unity of design, as shall enchain regard, ripen the instincts of his children to a finer sense of the bounties they enjoy, and kindle the admiration of every intelligent observer.

A neglect of these attractions, which are so conspicuous along all the by-ways of England, and in many portions of the continent, is attributable perhaps in some degree to the unrest of much of our rural population. The man who pitches his white tent beside the road, for what forage he may easily gather up, and is ready always for a sale, will care little for any of the more delicate graces of home. And with those who have some permanent establishment, I think the root of the difficulty may lie very much in that proud and sensitive individuality which is the growth of our democratic institutions. There is an absolute and charming fittingness about most of these humble rural adornments, of which I speak, which our progressive friend does not like to adopt, by reason of their *fittingness*, and because they give quasi indication of limited means and humble estate. When, therefore, such an one makes blundering effort to accomplish something in the way of decorative display, it is very apt to take a grandiose type, showing vulgar strain toward those adornments of the town which are wholly unsuited to his habits and surroundings. Thus a thriving ruralist with a family of two, will build a house as large as a church, and perch a cupola upon it, from which he may review the flat country for miles, while he contents himself with occupancy of the back-kitchen. If contented with small space, why not, in the name of honesty, declare it boldly, instead of covering the truth, under such lumbering falsehood? What forbids giving to the country home a simple propriety of its own, with its own wealth of rural decoration—

its shrubbery, its vines, its arbors, instead of challenging unfavorable comparison with an entirely different class of homes? If a man is disposed to advertise by flaming architecture and appointments—'I am only farmer by accident, and competent (as you see) to live in a grand way,' there is little hope that he will ever do anything to the credit of farming interests, or contribute very largely to the best charms of our rural landscape. The attempt to better one's condition is always praiseworthy; but it is only base and ignoble to attempt to cover one's condition with an idle smack of something larger.

There will always be in every moderately free country a great class of small landholders, in whose hands will lie for the most part, the control of our rural landscape, and the fashioning of our wayside homes, and when they shall take pride, as a body, in giving grace to these homes, the country will have taken a long step forward in the refinements of civilization. If I have no coaches and horses, I can at least hang a tracery of vine leaves along my porch, so exquisitely delicate that no sculpture can match it; if I have no conservatories with their wonders, yet the sun and I together can build up a little tangled coppice of blooming things in my door-yard, of which every tiny floral leaflet shall be a miracle. Nay, I may make my home, however small it be, so complete in its simplicity, so fitted to its offices, so governed by neatness, so embowered by wealth of leaf and flower, that no riches in the world could add to it, without damaging its rural grace; and my gardeners—Sunshine, Frost, and Showers are their names—shall work for me with no crusty reluctance, but with an abandon and a zeal that ask only gratitude for pay.

But let us come to details.

Walks.

A WALK is, first of all, a convenience; whether leading from door to highway, or to the stable court, or through gardens, or to the wood, it is essentially, and most of all—a convenience; and to despoil it of this quality, by interposing circles or curves, which have no meaning or sufficient cause, is mere affectation. Not to say, however, that all paths should be straight; the farmer, whose home is at a considerable remove from the highway, and who drives his team thither, avoiding rock, and tree, and hillock, will give to his line of approach a grace that it would be hard to excel by counterfeit. Pat, staggering from the orchard, under a bushel of Bartlett pears, and seizing upon every accidental aid in the surface of the declivity to relieve the fatigue of his walk—zigzagging, as it were in easy curves, is unconsciously laying down—though not a graceful man—a very graceful line of march. And it is the delicate interpretation of these every-day deflexities, and this instinctive tortuousness (if I may so say), which supplies, or should supply, the landscape gardeners with their best formulæ.

There is no liver in the country so practical, or of so humble estate, but he will have his half dozen paths divergent from his door; and these he may keep dry, and in always serviceable condition, by simply removing the soil from them to the depth of eighteen or twenty inches, and burying in them the scattered stones and debris, which are feeding weed-crops in idle corners; he will thus relieve himself of the useless material that might cumber the highway, besides possessing himself of the greater part of the top soil removed, for admixture with his composts. And this substitution may progress, season by season; as the garden rakings or refuse material accumulate, he has only to remove a few cubic yards of earth from his paths, bury the waste, and reserve the more available portions of the mould.

The same rules of construction are good for all road-ways, more especially for the farmer who wants unyielding metal beneath his heavy cartage of spring. The perfection of roads of course supposes perfect drainage, and a deep bed of stone material; but I am only suggesting methods which are in keeping with ordinary farm economies.

There must needs be directness in all paths communicating with out-buildings, and the exigencies of economic and effective culture demand the straight lines in the kitchen garden; but when I take a friend to some pretty point of view, or a little parterre of flowers dropped in the turf,—we are not hurried; the dainty curves make a pleasant cheatery of the approach. Thus there is charming accord between the best rules for landscape outlay, and the wants of the country-liver; where economy of tillage or of labor demands directness, the paths should be direct; and where economy of pleasure suggests loitering, the paths may loiter. And so, they loiter away through pleasant wooded coppices—doubling upon themselves on some rocky pitch of hill—short reaches, concealed each one from the other—blinded by thick underwood—wantoning in curves, until presently from under a low-branching beech tree, there bursts on the eye a great view of farm, and forest, and city, and sea; always a charming view indeed, though we toiled straight toward it, in broad sunshine; but the winding through the coppice, unsuspecting,—busied with ferns and lichens, and shut in by dark overgrowth against any glimpse of sky,—makes it tenfold ravishing.

What if such walks be not nicely gravelled—what if you come upon no grubbing gardeners? If only they be easy and serviceable, I love their rain stains, and their fine mosses creeping into green mats; I love their irregular borders, with a fern or a gentian nodding over the bounds—a pretty sylvan welcome to your tread. There are little foot-paths I know,—only beaten by the patter of young feet—winding away through lawn or orchard to some favorite apple tree,—frequented most, after some brisk wind-storm has passed over,—that I think I admire more than any gravelled walks in the world.

And there are other simple foot-paths, which I remember loitering through day after day, in the rural districts of England, with a sense of enjoyment, that never belonged to saunterings in the alleys of Versailles.

A man does not know England, or English landscape, or English country feeling, until he has broken away from railways, from cities, from towns, and clambered over stiles, and lost himself in the fields.

—— Talk of Chatsworth, and Blenheim, and Eaton Hall! Does a man know the pleasure of healthy digestion by eating whip-syllabub? Did Turner go to Belvoir Castle park for the landscapes which link us to God's earth?

What a joy and a delight in those field foot-paths

of England! Not the paths of owners only; not cautiously gravelled walks; but all men's paths, where any wayfarer may go; worn smooth by poor feet and rich feet, idle feet and working feet; open across the fields from time immemorial; God's paths for his people, which no man may shut;—winding—coiling over stiles—leaping on stepping-stones through brooks—with curves more graceful than Hogarth's—hieroglyphics of the Great Master written on the land, which, being interpreted, say—Love one another.

We call ours a country of privilege, yet what rich man gives right of way over his grounds? What foot-path or stile to cheat the laborer of his fatigue?

Shrubbery.

DOES the reader remember that upon the June day on which I first visited My Farm, I described the air as all aflow with the perfume of purple lilacs; and does he think that I would ungratefully forget it, or forget the lilacs? The Lilac is one of those old shrubs which I cling to with an admiration that is almost reverence. The Syringo (Philadelphus) is another; and the Guelder-rose (Viburnum) is another. They are all infamously common; but so is sunshine.

The Mezereum, the Forsythia, and the Weigelia

have their attractions ;—the Mezereum, because it is first comer in the spring, and shows its modest crimson tufts of blossoms, while the March snows are lingering ; the Forsythia follows hard upon it, with its graceful yellow bells ; and the Weigelia, though far later, is gorgeous in its pink and white—but neither of them is to be matched against the old favorites I have named.

Yet it is after all more in the disposition of the shrubbery, than in the varieties, that a rational pleasure will be found. It is not a great burden of bloom from any particular shrub that I aim at. I do not want to prove what it may do at its best, and singly ; that is the office of the nurseryman, who has his sales to make. But I want to marry together great ranks of individual beauties, so that May flowers shall hardly be upon the wane, when the blossoms of June shall flame over their heads ; and June in its turn have hardly lost its miracles of color, when July shall commence its intermittent fires, and light up its trail of splendor around all the skirts of the shrubbery. I want to see the delicate white of the Clematis (Virginica) hanging its graceful festoons of August, here and there in the thickets that have lost their summer flowers ; and after this I welcome the black berries of the Privet, or the brazen ones of the twining Bitter-sweet.

Or, it is some larger group with which we deal—half up the hill-side, screening some ragged nursery of rocks—and a tall Lombardy-poplar lifts from its centre, while shining, yellowish Beeches group around it—crowding it, forcing all its leafy vigor (just where we wish it) into the topmost shoots; and amid the Beeches are dark spots of young Hemlocks—as if the shadow of a cloud lay just there, and the sun shone on all the rest; and among the Hemlocks, and reaching in jagged bays above and below them are Sumacs (so beautiful, and yet so scorned) lifting out from all the tossing sea of leaves, their solid flame jets of fiery crimson berries. Skirting these, and shining under the dip of a Willow, are the glossy Kalmias which, at midsummer, were a sheet of blossoms; and the hem of the group is stitched in at last with purple Phloxes and gorgeous Golden-rod.

I know no limit indeed to the combinations which a man may not affect who has an eye for color, and a heart for the light labor of the culture. There is, unfortunately, a certain stereotyped way of limiting these shrubberies to a few graceful exotics,—which, of course, the gardeners commend,—and of rating the value of foliage by its cost in the nursery. It is but a narrow and ungrateful way of dealing with the bounties of Providence. It may accomplish under great care, very effective results; but they will not open the eyes of men of humble estates to the beauties that are lurking in the forest all around them, and which only need a little humanizing care to rival the best products of the nurseries. Steering clear of this intolerance, I have domesticated the White-birch, and its milky bole is without a rival among all the exotics; the Hardbeam (Carpinus), with its fine spray, and the Witch-hazel (Hamamelis virginica) with its unique bloom upon the bare twigs of November, are thriving in my thickets. The swamp Azalias, and the Kalmias I have transferred successfully, in their season of flowering.* There are also to be named among the available native shrubs,—the Leather-wood (Dirca palustris) with delicate yellow bloom, glossy green leaves, and an amazing flexibility of bough, on which once a year my boy forages for his whip-lashes; the Spice-wood (Laurus benzoin) is always tempting to the children by reason of its aromatic bark, and in earliest spring it is covered with fairy golden flowers; the spotted Alder is a modest shrub through the summer, but in autumn it flames out in a great harvest of scarlet berries, which it carries proudly into the chills of December; the red-barked Dog-wood (Cornus sanguinea) supplies annually a great stock of crimson whips, and a charming liveli-

* A much safer way is to give the young plants a season or two of domestication in a patch of nursery ground.

ness of color for any interior rustic ornamentation, which a wet day may put in hand; the Swamp-willow is the very earliest of our native shrubs, to feel the heats of the March sun, and season after season, the little ones bring in from its clump, its silvery strange tufts of bloom, and say: "The Willow mice have come,—and the spring."

Nor must I forget the Barberry, beautiful in its bloom, and still more beautiful with its crimson fruit, —the May-flower, the Sumac, the Sweet-brier, the Bilberry, with its fairy bells, and the whole race of wild vines—among which not least, is the luxuriant Frost-grape, tossing its tendrils with forest freedom from the tops of the tallest trees, and in later June filling the whole air with the exquisite perfume of its blossoms.

It may seem that a great estate and wide reach of land may be demanded for the aggregation of all these denizens of the wood; yet it is not so; I have all these and more than these, with room for their own riotous luxuriance, in scattered groups and copses, without abstracting so much as an acre from the tillable surface of the land. The brambles, thickets, and unkempt hedge-rows which half the farmers of the country leave to encroach upon the fertility and order of their fields, work tenfold more of harm, than the coppices which I have planted on rocky declivities, and on lands, else unserviceable; or as a shelter to my garden or poultry yard,—as a screen from the too curious eyes of the public;—tangled wildernesses, not without an order of their own,—offering types of all the forest growth, where the little ones may learn the forest names, and habit—a living book of botany, whose tender lessons are read and remembered, as the successive seasons waft us their bloom and perfume.

These groups will, of course, demand some care for their effective establishment; care is a price we must all pay for whatever beautiful growth we secure—whether in our trees or our lives.

It is specially imperative that all turf be removed, wherever a group of shrubs or forest trees are to be planted; trenching is by no means essential, and with many of the forest denizens, promotes a woody luxuriance that delays bloom. My own practice has been to compost the turf as it was taken up, upon the ground, with lime, and possibly a castor-pomace, or other nitrogenous fertilizer; this I reserved for a top-dressing, as the shrubs might seem to require, and no other application of manure is ever made. Three times, the first year, and twice, the second year, it may be necessary to give hoe-culture, in order to keep the grass and other foreign growth in abeyance. After this, a single dressing is amply sufficient; and on his after-dinner strolls to the thickets, the planter will not forget his pruning-knife and his saw.

A little patch of good, and thoroughly tilled nursery-ground is very convenient as a tender upon these wood-groups, as well as upon the orchard. Within a small one of my own—of less than an eighth of an acre, I have now thriving hundreds of hemlocks, white-pines, birches, maples, alders, vines, beeches, willows, kalmias,—with which I may at any time thicken up the skirts of the established groups to any color I like, or plant a new one upon some scurvy bit of land, which has proved itself unremunerative under other croppings.

Altogether, these shows of forest foliage, with here and there an exotic, or a fruit-tree thrown in,—involve less cost than one would give to an ordinary crop of corn; and when the corn is harvested, the crop is done; but with my shrubberies—of which I know every tree from the day of its first struggle with the changed position—the weird, wild growth is every year progressing—every year presenting some new phase of color or of shape:—every spring I see my trees rejoicing in a flutter of young leaves, and then wantoning—like grown girls—in the lusty vigor of summer: in autumn I look wistfully on them, wearing gala-dresses, whose colors I dare not name, and when these are shivered by the frost,—

tranquilly disrobing, and retiring to the sleep of winter.

Rural Decoration.

AMONG the things which specially contribute to the charms of a country-home, are those thousand little adornments, which a person of quick observation and ready tact can easily avail himself of; and while gratifying his own artistic perceptions, he can contribute to the growth of a humble art-love, which it is to be hoped will some day give a charm to every road-side, and to every country cottage. It is by no means true that a taste of this kind must necessarily—like Sir Visto's—prove a man's ruin. The land is indeed a great absorbent; and if no discretion be brought to the direction of outlay in adornments and improvements, or if they be not ordered by a severe and inexorable simplicity, it is quite incredible what amounts of money may be expended.

I have in an earlier portion of this volume, hinted at certain changes which may be made, in the throwing out of some half dozen angular and unimportant enclosures, at the door, into open lawn—in the removal of unnecessary fences, and the establishment of groups of shrubbery to hide roughness, or to furnish shelter: all which involve little expenditure, and are not in violation of any rules of well-considered economy. I may now add to these the effects of little unimportant architectural devices, not requiring a practical builder, and which while they lend a great charm to landscape, give an individuality to a man's home.

The reader will perhaps allow me to particularize from my own experience. There were, to begin with, some four or five disorderly buildings about the farm-house—sheds, shops, coal-houses, smoke-houses—built up of odds and ends of lumber—boards matching oddly, some half painted, others too rough for paint—altogether scarcely bad enough for removal, and yet terribly slatternly and dismal in their general effect. They were not worth new covering; painting was impossible; and whitewashing would only have lighted up the seams and inequalities more staringly. A half a mile away was a little mill, where cedar posts were squared by a circular saw, and the slabs were packed away for fuel (and very poor fuel they made). One day, as my eye lighted upon them, an idea for their conversion to other uses struck me, and fructified at once. I bought a cord or two at a nominal rate, and commenced the work of covering my disjointed and slatternly outbuildings with these rough slabs. It was a simple business, requiring only even nailing, with here and there a little 'furring out' to bring the old angles to a square, with here and there the deft turning of a rude arch, with two crooked bits, over door or window. Farm laborers, under direction, were fully competent to the work; and in a couple of days I had converted my unsightly buildings into very tasteful, rustic affairs, harmonizing with the banks of foliage behind and over them, and giving capital foothold to the vines which I planted around them.

In keeping with their effect, I caused gates to be constructed of the simplest material, from the cedar thickets; varying these in design, and yet making each so simple as to admit of easy imitation, and to unite strength, solidity, and cheapness. If, indeed, these latter qualities could not be united, the work would not at all meet the end I had in view—which was not merely to produce a pretty effect, but to demonstrate the harmony of such decorative work with true farm economy. One often sees, indeed, rustic-work of most cumbrous and portentous dimensions —overladen with extraordinary crooks and curves, and showing at a glance immense labor in selection and in arrangement. All this may be pleasing, and often exceedingly beautiful; but it is a mere affectation of rural simplicity; it wears none of that fit and simple character which would at once commend it to the eye of a practical man as an available and imitable feature. If I can give such arrangement to simple boughs, otherwise worthless, or to pine-pickets of simple cost—in the paling of a yard, or the tracery of a gate, as shall catch the eye by its grace of outline, and suggest imitation by its easy construction, and entire feasibility, there is some hope of leading country tastes in that direction; but if work shows great nicety of construction, puzzling and complicated detail, immense absorption of labor and material, it might as well have been—so far as intended to encourage farm ruralities—built of Carrara marble.

Again a stone wall, or dyke, is not generally counted an object of much beauty, except it be laid up in hammered work; this, of course, is out of the question for a farmer who studies economy: but suppose that to a substantial stone fence of ordinary construction, I am careful, by a choice of topping-stones, to give unbroken continuity to its upper line; and suppose that the abutments, instead of wearing the usual form, are carried up a foot or more above this line in a rude square column, gradually tapering or 'battering' toward the top; suppose upon this top I place a flat stone nearly covering it, and upon this a smaller stone some four inches in thickness, and again, upon the last, the largest and roundest boulder I can find. At once there is created a graceful architectural effect, which gives a new air to the whole line of wall. Yet the additional labor involved is hardly to be reckoned.

Gates, in all variety, dependent on position and service, offer charming opportunity for simple and effective rural devices. Far away in the garden it may be worth while to throw a rude rooflet over one, where a man may catch refuge from a shower; in another quarter, you may carry up posts and link them across in rustic trellis, to carry the arms of some tossing vine; a stile, too, where neighbors' children, forgetful of latches, are apt to stroll in for nuts or berries, or on some cross-path to school, may, by simple adjustment of log steps and overhanging roof of thatch, or slabs, take a charming effect, and work somewhat toward the correction of that unflinching and inexorable insistance upon rights of property, which induces many a crabbed man to nail up his gates, and deny himself a convenience, for the sake of circumventing the claims of an occasional stroller.

Rustic seats are an old and very common device; but with these, as with gateways and palings, simplicity of construction is the grand essential. I see them not unfrequently so fine and elaborate, that one fears a shower may harm them; and when so fine as to suggest this fear, they had much better be of rosewood and bamboo. A simple bit of plank between two hoary trunks—held firmly in place by the few bits of gnarled oak-limbs from which arms, legs, and back are adroitly—hinted, rather than fashioned—is more agreeable to country landscape, fuller far of service and of suggestion, than any of the portentous rustic-work in city shops.

The due adjustment of colors is also a thing to be considered in the reckoning of rural effects; thus, with my old weather-stained house, I do not care to place new paint in contrast; the old be-clouded tint harmonizes well with the rustic work of fences and outbuildings; while away, upon the lawn, or opening into green fields, or—better still—in the very bight of the wood, I give the contrast of a brilliant and flashing white.

I am touching a very large subject here, with a very short chapter. Indeed, there is no end to the pretty and artistic combinations by which a man who loves the country with a fearless, demonstrative love, may not provoke its rarer beauties to appear. Flower, tree, fence, outbuilding—all wait upon his hands; and the results of his loving labor do not end when his work is done; but the vines, the trees, the mosses, the deepening shadows, are, year after year, mellowing his raw handiwork, and ripening a new harvest of charms. And in following these, I think there is an interest—not perhaps quotable on 'Change, but which rallies a man's finer instincts, and binds him in leash—not wearisome or galling—to the great procession of the seasons, ever full of bounties, as of beauties.

Flowers.

THERE is a class of men who gravitate to the country by a pure necessity of their nature; who have such ineradicable love for springing grass, and fields, and woods, as to draw them irresistibly into companionship. Such men feel the confinement of a city like a prison. They are restive under its restraint. The grass of an area patch of greensward kindles their love into flame. They linger by florists' doors, drawn and held by a magnetism they cannot explain, and which they make no effort to resist. They are not necessarily amateurs, in the ordinary sense of that term. I think they are apt to be passionate lovers of only a few, and those the commonest flowers—flowers whose sweeet home-names reach a key, at whose touch all their sympathies respond.

They laugh at the florist's fondness for a well rounded holly-hock, or a true-petalled tulip, and admire as fondly the half-developed specimens, the careless growth of cast-away plants, or the accidental thrust of some misshapen bud or bulb. I suspect I am to be ranked with these; my purchase of an ox-eye daisy on the streets of Paris will have already damaged my reputation past hope, in the eyes of the amateur florists. If these good people could see the homely company of plants that is gathered every winter in my library window, they would be shocked still farther.

There is a careless group of the most common ferns; a Rose-geranium, a Daphne, a common Monthly-rose, are the rarest plants I boast of. But there are wood-mosses with a green sheen of velvet; they cover a broad tray of earth in rustic frame-work, in which the Geraniums, the mosses, the Daphne, and a plant of Kenilworth-Ivy coquette together. An upper shelf is embossed with other mosses; there is a stately Hyacinth or two lifting from among them, and wild ferns hang down their leaves for a careless tangle with the Geraniums and Ivy below. Above all, and as a drapery for the arched top, the Spanish moss hangs like a gray curtain of silvered lace.

A stray acorn, I observe, has shot up in the tray, and is now in its third leaf of oak-hood; in the corners, two wee Hemlock-spruces give a background of green, and an air of deeper and wilder entanglement, to my little winter-garden. A bark covering, with bosses of acorn-cups, and pilasters of laurel-wood, sharpened to a point, make the lower tray a field of wildness,—fenced in with wildness. The overhanging bridge (I called it an upper shelf) is a rustic gallery —its balcony of twisted osiers filled in with white mosses from old tree-stumps, and the whole supported by a rustic arch of crooked oaken twigs. Finally, the cornice from which the Spanish moss is pendant, is a long rod of Hazel, around which a vine of Bitter-sweet has twined itself so firmly, that they seem incorporate together; and to their rough bark the moss has taken so kindly, that it has bloomed two full years after the date of its first occupancy. There are daintier hands than mine that care for this little garden of wildness, and give it its crowning grace; but here—I may not speak their praise.

The other southern window is at a farther remove from the open wood-fire; its floral show is, therefore, somewhat different; and the reader will, I trust, excuse me a little particularity of description, since it will enable me to show how much may be done with limited material and space.

Upon the window-sill,—some eighteen inches in breadth by forty in length,—are placed four bits of oak-wood five inches in length, squarely sawn from a young forest tree, which serve as standards or supports, to a tray of plank five inches in depth, and covered with unbarked saplings, so graduated in size, as to make this base (or tray) appear like the plinth of a column. This is filled with fine garden-mould, and there are grooves in the plank-bottom communicating with one drainage hole, beneath which is placed an earthen saucer. Fitting upon this tray is a glazed case with top sloping to the sun, and with its quoins and edges covered with bark, and embossed with acorn-cups—to correspond with the base. The fitting is not altogether so perfect as that of a Wardian case, but quite sufficient for all practical purposes.

Throughout the summer I keep this little window-garden stocked with the most brilliant of the wood mosses; a slight sprinkling once in thirty days keeps them in admirable order; and if I come upon some chrysalis in the garden whose family is unknown, I have only to lodge it upon my bed of mosses, and in due time I have a butter-fly captive for further examination. As the frosts approach I throw out my mosses, and re-stock my garden with fragrant violets and a few ferns. These keep up a lusty garden show until January, when again I change the order of my captives—this time incorporating a large share of sand with the earth in the tray—and setting in it all my needed cuttings of Verbenas, of Fuchsias, and of Carnations. They thrive under the glass magically; and by early March are so firm-rooted and rampant in growth, that I can pot them, for transfer to a fresh-laid pit out of doors. I now amend the soil, and sprinkling it with a dash of ammoniacal water, sow in it the Cockscomb, Peppers, Egg-plants, and whatever fastidious plants require special care, while along the edges I prove my over-kept cabbage and clover seed. All these make their way, and in due time come to their season of potting, when I give up my little garden to a careless array of the first laughing flowers of spring.

Can you tell me of so small a window anywhere that shows so many stages of growth? Nor have I named all even yet. A rustic arch, steep as the Rialto at Venice, overleaps this tiny garden, and bears upon its centre a miniature Swiss chalet, while down either flank, upon successive steps, are little bronze mementos of travel—among which the delicate tendrils of a German-ivy (planted upon a ledge of its own) intertwine and toss their tender leaflets into the doors and windows of the chalet.

But I am lingering in-doors, when my book is essentially an out-of-door book.

I am not about to lay down any rules for flower-beds or for flower culture; the gardening books are full of them; and by their aid, and that of a dexterous gardener, any one may arrange his parterres and his graduated banks of flowers, quite *secundum artem.* And I suppose, that, when completed, these orderly arrays of the latest and newest floral wonders are enjoyable. Yet I am no fair judge; the appreciation of them demands a 'booking-up' in floral science to which I can lay no claim. I sometimes wander through the elegant gardens of my town friends, fairly dazzled by all the splendor and the orderly ranks of beauties; but nine times in ten—if I do not guard my tongue with a prudent reticence, and allow my admiration to ooze out only in exclamations—I mortify the gardener by admiring some timid flower, which nestles under cover of the flaunting Dahlias or Peonies, and which proves to be only some dainty weed, or an antiquated plant, which the florists no longer catalogue. Everybody knows how ridiculous it is to admire a picture by an unknown artist; and I must confess to feeling the fear of a kindred ridicule, whenever I stroll through the gardens of an accomplished amateur.

But I console myself with thinking that I have company in my mal-adroitness, and that there is a great crowd of people in the world, who admire spontaneously what seems to be beautiful, without waiting for the story of its beauty. If I were an adept, I should doubtless, like other adepts, reserve my admiration exclusively for floral perfection; but I thank God that my eye is not as yet so bounded. The blazing Daffodils, Blue-bells, English-cowslips, and Striped-grass, with which some pains-taking woman in an up-country niche of home, spots her little door-yard in April, have won upon me before now to a tender recognition of the true mission of flowers, as no gorgeous parterre could do.

With such heretical views, the reader will not be

surprised if I have praises and a weakness for the commonest of flowers. Every morning in August, from my chamber window, I see a great company of the purple Convolvulus, writhing and twisting, and over-running their rude trellis, while above and below, and on either flank of the wild arbor, their fairy chalices are beaded with the dew. A Scarlet-runner is lost—so far as its greenness goes—in the tangle of a hedge-row, and thrusts out its little candelabras of red and white into the highway, to puzzle the passers-by, who admire it,—because they do not know it. A sturdy growth of Nasturtium is rioting around the angle of a distant mossy wall, at the end of a woody copse—so far away from all parterres, that it seems to passers some strange, gorgeous wild-flower; and yet its blaze of orange and crimson is as common and vulgar as the wood-fire upon a farmer's hearth. Holly-hocks—so far away you cannot tell if they be double or single (they are all single)—lift their stately yellows and whites in the edge of the shrubbery; Phloxes, purple and white, hem them in; and at their season a wilderness of Roses bloom in the tangled thicket.

Dotted about here and there, in unexpected places —yet places where their color will shine—are clumps of yellow Lilies, of Sweet-William, of crimson Peonies, of Larkspur, or even (shall I be ashamed to tell it?)

of Golden-rod and of the Cardinal flower (Lobelia). In a little bed scooped from the turf and bordering upon the nearer home-walks, are the old-fashioned Spider-wort, and that old white Lily, which Raphael makes the Virgin hold on the day of her espousals. And yet you may go through half the finest gardens of the country and never find this antiquated Lily! The sweet Violet and the Mignonette have their place in these new borders, as well as the roses. Cypress and Madeira vines twine, in leash with the German ivy, over a pile of stumps that have been brought down from the pasture; under the lee of a thicket of pines, among lichened stones heaped together, is a group of ferns and Lycopodiums; and the sweet Lily of the Valley,—true to its nature and quality,—thrives in a dark bit of ground half shaded between two spurs of a bushy thicket.

Of course, there are the Verbenas, for which every year a fresh circlet of ground is prepared from the turf, and a great tribe of Geraniums, to bandy scarlets with the Salvias; and the Fuchsias, too—though very likely not the latest named varieties; nor are they petted into an isolated, pagoda-like show, but massed together in a little group below the edge of the fountain, where they will catch its spray, and where their odorless censers of purple and white and crimson may swing, or idle, as they will. And among the mossy stones from amid which the fountain gurgles over, I find lodging places, not only for rampant wild-ferns, but for a stately Calla, and for some showy type of the Amaryllidæ.

It is in scattered and unexpected places, that I like my children to ferret out the wild-flowers brought down from the woods—the frail Colombine in its own cleft of rock,—the Wild-turnip, with its quaint green flower in some dark nook, that is like its home in the forest—the Maiden's-hair thriving in the moist shadow of rocks; and among these transplanted wild ones of the flower-fold, I like to drop such modest citizens of the tame country as a tuft of Violets, or a green phalanx of the bristling Lilies of the Valley.

Year by year, as we loiter among them, after the flowering season is over, we change their *habitat*, from a shade that has grown too dense, to some summer bay of the coppices; and with the next year of bloom, the little ones come in with marvellous reports of Lilies, where Lilies were never seen before —or of fragrant Violets, all in flower, upon the farthest skirt of the hill-side. It is very absurd, of course; but I think I enjoy this more—and the rare intelligence which the little ones bring in with their flashing, eager eyes—than if the most gentlemanly gardener from Thorburn's were to show a Dahlia, with petals as regular as if they were notched by the file of a sawyer.

Flowers and children are of near kin, and too much of restraint or too much of forcing, or too much of display, ruins their chiefest charms. I love to associate them, and to win the children to a love of the flowers. Some day they tell me that a Violet or a tuft of Lilies is dead; but on a spring morning, they come, radiant with the story,—that the very same Violet is blooming sweeter than ever, upon some far away cleft of the hill-side. So you, my child, if the great Master lifts you from us, shall bloom—as God is good—on some richer, sunnier ground!

——We talk thus: but if the change really come, it is more grievous than the blight of a thousand flowers. She, who loved their search among the thickets—will never search them. She, whose glad eyes would have opened in pleasant bewilderment upon some bold change of shrubbery or of paths, will never open them again. She—whose feet would have danced along the new wood-path, carrying joy and merriment into its shady depths,—will never set foot upon these walks again.

What matter how the brambles grow?—her dress will not be torn: what matter the broken paling by the water?—she will never topple over from the bank. The hatchet may be hung from a lower nail now—the little hand that might have stolen possession of it, is stiff—is fast! God has it.

And when spring wakens all its echoes—of the wren's song—of the blue-bird's warble,—of the plaintive cry of mistress cuckoo (*she* daintily called her mistress cuckoo) from the edge of the wood—what eager, earnest, delighted listeners have we—lifting the blue eyes,—shaking back the curls—dancing to the melody? And when the violets repeat the sweet lesson they learned last year of the sun and of the warmth, and bring their fragrant blue petals forth—who shall give the rejoicing welcome, and be the swift and light-footed herald of the flowers? Who shall gather them with the light fingers, she put to the task—who?

And the sweetest flowers wither, and the sweetest flowers wait—for the dainty fingers that shall pluck them, never again!

L'Envoi.

I HAVE now completed the task which I had assigned to myself; and I do it with the burdensome conviction that not one half of the questions which suggest themselves in connection with Farm-life in America, can be discussed—much less resolved—within so narrow a compass. Yet I have endeavored to light up, with my somewhat disorderly array of hints and suggestions, those more salient topics which would naturally suggest themselves to all who may have a rural life in prospect, or who may to-day be idling or planning, or toiling under the shadow of their own trees.

There are no grand rules by which we may lay down the proportions of a life, or the wisdom of this or that pursuit; every man is linked to his world of duties by capacities, opportunities, weaknesses, which will more or less constrain his choice. And I am slow to believe that a man who brings cultivation, refinement, and even scientific attainment, may not find fit office for all of them in country life, and so dignify that great pursuit in which, by the necessity of the case, the majority of the world must be always engaged. He may contribute to redeem it from those loose, immethodical, ignorant practices, which are, in a large sense, due to the farmer's isolation, and to the necessities of his condition. And although careful investigation, study, and extended observation in connection with husbandry, may fail of those pecuniary rewards, which seem to be their due, yet the cause in some measure ennobles the sacrifice. The cultivated farmer is leading a regiment in the great army whose foraging success is feeding the world; and if he put down within the sphere of his influence—riotous pillage—wasteful excesses, and by his example give credit to order, discipline, and the best graces of manhood,—he is reaping honors that will endure :—not measured by the skulls he piles on any Bagdad plains, but by the mouths he has fed—by the flowers he has taught to bloom, and by the swelling tide of harvests which, year by year, he has pushed farther and farther up the flanks of the hills.

I would not have my reader believe that I have carried out as yet within the limits of the farm herein described all that I have advised—whether in the things which relate to its productive capacity, or to its embellishment. All this ripens by slow progression which we cannot unduly hasten. Nor do I know that full accomplishment would add to the charm; I think that those who entertain the most keen enjoyment of a country homestead, are they who regard it always in the light of an unfinished picture—to which, season by season, they add their little touches, or their broad, bold dashes of color; and yet with a vivid and exquisite foresight of the future completed charm, beaming through their disorderly masses of pigments, like the slow unfolding of a summer's day.

In all art, it is not so much the bald image that meets the eye, as it is the crowd of suggested images lying behind, and giving gallant chase to our fancy—which gives pleasure. It is not the mere palaces in the picture of Venice before my eye, which delight me, but the reach of imagination behind and back of them—the shadowy procession of Doges—the gold cloth—the Bucintoro—the plash of green water kissing the marble steps, where the weeds of the Adriatic hang their tresses, and the dainty feet of Jessica go tripping from hall to gondola. It is not the shaggy, Highland cattle, with dewy nostrils lifted to the morning, that keep my regard in Rosa Bonheur;—but the aroma of the heather, and of a hundred Highland traditions,—a sound—as of Bruar water,—a sudden waking of all mountain memories and solitudes.

Again it must be remembered by all those who have rural life in anticipation, that its finer charms, and those which grow out of the adornments and accessories of home, are dependent much more upon the appreciative eye and taste of the mistress than of the master. If I have used the first person somewhat freely in my descriptions, it has been from no oversight of what is justly due to another; and I would have the reader believe—what is true—that all the more delicate graces which are set forth, and which spring from flowers or flowering shrubs, and their adroit disposition, are due to tenderer hands, and a more provident and appreciative eye than mine.

I think that I have not withheld from view the awkwardnesses and embarrassments which beset a country life in New England,—nor overstated its possible attractions. I have sought at any rate, to give a truthful picture, and to suffuse it all—so far as I might—with a country atmosphere; so that a man might read, as if the trees were shaking their leaves over his head,—the corn rustling through all its ranks within hearing, and the flowers blooming at his elbow.

Be this all as it may,—when, upon this charming morning of later August, I catch sight, from my window, of the distant water—where, as at the first—white sails come and go:—of the spires and belfries of the near city rising out of their bower of elms—of the farm lands freshened by late rains into unwonted greenness;—of the coppices I have planted, shaking their silver leaves, and see the low fire of border flowers flaming round their skirts, and hear the water plashing at the door in its rocky pool, and the cheery voices of children, rejoicing in health and the country air,—I do not for a moment regret the first sight of the old farm house, under whose low-browed ceiling, I give this finishing touch to the last chapter of MY FARM OF EDGEWOOD.

www.ingramcontent.com/pod-product-compliance
Lightning Source LLC
LaVergne TN
LVHW020215110826
845151LV00003B/725

9781425533984